DISEASES OF HORTICULTURAL CROPS

Diagnosis and Management

Volume 4: Important Plantation Crops, Medicinal Crops, and Mushrooms

Diseases of Horticultural Crops: Diagnosis and Management, 4 Volume Set:

Volume 1: Fruit Crops

Volume 2: Vegetable Crops

Volume 3: Ornamental Plants and Spice Crops

Volume 4: Important Plantation Crops, Medicinal Crops, and Mushrooms

Innovations in Horticultural Science

DISEASES OF HORTICULTURAL CROPS

Diagnosis and Management

Volume 4: Important Plantation Crops, Medicinal Crops, and Mushrooms

Edited by

J. N. Srivastava, PhD

A. K. Singh, PhD

First edition published 2022

Apple Academic Press Inc.
1265 Goldenrod Circle, NE,
Palm Bay, FL 32905 USA
4164 Lakeshore Road, Burlington,
ON, L7L 1A4 Canada

CRC Press
6000 Broken Sound Parkway NW,
Suite 300, Boca Raton, FL 33487-2742 USA
2 Park Square, Milton Park,
Abingdon, Oxon, OX14 4RN UK

Library and Archives Canada Cataloguing in Publication

Title: Diseases of horticultural crops : diagnosis and management / edited by J.N. Srivastava, PhD, A.K. Singh, PhD.

Names: Srivastava, J. N. (Plant pathologist), editor. | Singh, A. K. (Plant pathologist), editor.

Series: Innovations in horticultural science.

Description: First edition. | Series statement: Innovations in horticultural science | Includes bibliographical references and indexes. | Content: Volume 4: Important Plantation Crops, Medicinal Crops, and Mushrooms.

Identifiers: Canadiana (print) 20210324392 | Canadiana (ebook) 2021032452X | ISBN 9781771889926 (v. 4 ; hardcover) | ISBN 9781774639436 (v. 4 ; softcover) | ISBN 9781771889889 (set) | ISBN 9781003160472 (v. 4 ; ebook)

Subjects: LCSH: Plant diseases—Diagnosis. | LCSH: Phytopathogenic microorganisms—Control.

Classification: LCC SB731 .D57 2022 | DDC 632—dc23

Library of Congress Cataloging-in-Publication Data

Names: Singh, A. K. (Plant pathologist), editor. | Srivastava, J. N. (Plant pathologist), editor.

Title: Diseases of horticultural crops: diagnosis and management. Volume 4, Important plantation crops, medicinal crops, and mushrooms / J. N. Srivastava, A. K. Singh.

Other titles: Innovations in horticultural science.

Description: First edition. | Palm Bay, FL, USA : Apple Academic Press, 2022. | Series: Innovations in horticultural science | Includes bibliographical references and index. | Summary: "Diseases of Horticultural Crops: Diagnosis and Management: Volume 4: Important Plantation Crops, Medicinal Crops, and Mushrooms discusses the key diseases, typical symptoms, and management strategies of several economically important plants. Each chapter presents an introduction along with a detailed account of symptoms, causal organisms, disease cycles, epidemiology, and management of a selection of major plantation crops, medicinal crops, and mushrooms. The book features chapters contributed by eminent professionals in the field, who have incorporated their own experience and knowledge along with an overview of the recent development in their fields. They provide information on the diagnostic tools necessary and management techniques for such plantation crops as areca nut (or betel nut), cocoa (or chocolate), coconut, coffee, and tea; such medicinal crops as isabgol and senna; along with several kinds of mushrooms. The chapters cover key diseases, typical symptoms, and management strategies. The volumes also include photographs that show symptoms of important diseases, which are helpful in disease diagnosis. This volume is part of the 4-volume Diseases of Horticultural Crops: Diagnosis and Management. Other volumes focus on fruit crops, vegetable crops, and ornamental plants and spice crops. These volumes will be valuable to scientists and researchers, faculty and students, administrators and many others in the discipline of plant pathology but also in other areas of agriculture and allied subjects"-- Provided by publisher.

Identifiers: LCCN 2021051195 (print) | LCCN 2021051196 (ebook) | ISBN 9781771889926 (v. 4 ; hardback) | ISBN 9781774639436 (v. 4 ; paperback) | ISBN 9781003160472 (v. 4 ; ebook)

Subjects: LCSH: Plant diseases--Diagnosis. | Plant diseases.

Classification: LCC SB731 .D575 2022 (print) | LCC SB731 (ebook) | DDC 571.9/2--dc23/eng/20211027

LC record available at https://lccn.loc.gov/2021051195

LC ebook record available at https://lccn.loc.gov/2021051196

ISBN: 978-1-77188-992-6 (hbk)
ISBN: 978-1-77463-943-6 (pbk)
ISBN: 978-1-00316-047-2 (ebk)

Dedicated to

My beloved parents for blessing me
My Uncle (Mama)
Dr. H. K. Srivastava
For my successes in every sphere of life

INNOVATIONS IN HORTICULTURAL SCIENCE

Editor-in-Chief:

Dr. Mohammed Wasim Siddiqui, Assistant Professor-cum- Scientist
Bihar Agricultural University | www.bausabour.ac.in
Department of Food Science and Post-Harvest Technology
Sabour | Bhagalpur | Bihar | P. O. Box 813210 | INDIA
Contacts: (91) 9835502897
Email: wasim_serene@yahoo.com | wasim@appleacademicpress.com

The horticulture sector is considered as the most dynamic and sustainable segment of agriculture all over the world. It covers pre- and postharvest management of a wide spectrum of crops, including fruits and nuts, vegetables (including potatoes), flowering and aromatic plants, tuber crops, mushrooms, spices, plantation crops, edible bamboos etc. Shifting food pattern in wake of increasing income and health awareness of the populace has transformed horticulture into a vibrant commercial venture for the farming community all over the world.

It is a well-established fact that horticulture is one of the best options for improving the productivity of land, ensuring nutritional security for mankind and for sustaining the livelihood of the farming community worldwide. The world's populace is projected to be 9 billion by the year 2030, and the largest increase will be confined to the developing countries, where chronic food shortages and malnutrition already persist. This projected increase of population will certainly reduce the per capita availability of natural resources and may hinder the equilibrium and sustainability of agricultural systems due to overexploitation of natural resources, which will ultimately lead to more poverty, starvation, malnutrition, and higher food prices. The judicious utilization of natural resources is thus needed and must be addressed immediately.

Climate change is emerging as a major threat to the agriculture throughout the world as well. Surface temperatures of the earth have risen significantly over the past century, and the impact is most significant on agriculture. The rise in temperature enhances the rate of respiration, reduces cropping periods, advances ripening, and hastens crop maturity, which adversely affects crop productivity. Several climatic extremes such as droughts, floods, tropical cyclones, heavy precipitation events, hot extremes, and heat waves cause a negative impact on agriculture and are mainly caused and triggered by climate change.

In order to optimize the use of resources, hi-tech interventions like precision farming, which comprises temporal and spatial management of resources in horticulture, is essentially required. Infusion of technology for an efficient utilization of resources is intended for deriving higher crop productivity per unit of inputs. This would be possible only through deployment of modern hi-tech applications and precision farming methods. For improvement in crop production and returns to farmers, these technologies have to be widely spread and adopted. Considering the above-mentioned challenges of horticulturist and their expected role in ensuring food and nutritional security to mankind, a compilation of hi-tech cultivation techniques and postharvest management of horticultural crops is needed.

This book series, Innovations in Horticultural Science, is designed to address the need for advance knowledge for horticulture researchers and students. Moreover, the major advancements and developments in this subject area to be covered in this series would be beneficial to mankind.

Topics of interest include:

1. Importance of horticultural crops for livelihood
2. Dynamics in sustainable horticulture production
3. Precision horticulture for sustainability
4. Protected horticulture for sustainability
5. Classification of fruit, vegetables, flowers, and other horticultural crops
6. Nursery and orchard management
7. Propagation of horticultural crops
8. Rootstocks in fruit and vegetable production
9. Growth and development of horticultural crops
10. Horticultural plant physiology
11. Role of plant growth regulator in horticultural production
12. Nutrient and irrigation management
13. Fertigation in fruit and vegetables crops
14. High-density planting of fruit crops
15. Training and pruning of plants
16. Pollination management in horticultural crops
17. Organic crop production
18. Pest management dynamics for sustainable horticulture
19. Physiological disorders and their management
20. Biotic and abiotic stress management of fruit crops
21. Postharvest management of horticultural crops
22. Marketing strategies for horticultural crops
23. Climate change and sustainable horticulture
24. Molecular markers in horticultural science
25. Conventional and modern breeding approaches for quality improvement
26. Mushroom, bamboo, spices, medicinal, and plantation crop production

BOOKS IN THE SERIES

- **Spices: Agrotechniques for Quality Produce**
  Amit Baran Sharangi, PhD, S. Datta, PhD, and Prahlad Deb, PhD
- **Sustainable Horticulture, Volume 1: Diversity, Production, and Crop Improvement**
  Editors: Debashis Mandal, PhD, Amritesh C. Shukla, PhD, and Mohammed Wasim Siddiqui, PhD
- **Sustainable Horticulture, Volume 2: Food, Health, and Nutrition**
  Editors: Debashis Mandal, PhD, Amritesh C. Shukla, PhD, and Mohammed Wasim Siddiqui, PhD
- **Underexploited Spice Crops: Present Status, Agrotechnology, and Future Research Directions**
  Amit Baran Sharangi, PhD, Pemba H. Bhutia, Akkabathula Chandini Raj, and Majjiga Sreenivas
- **The Vegetable Pathosystem: Ecology, Disease Mechanism, and Management**
  Editors: Mohammad Ansar, PhD, and Abhijeet Ghatak, PhD
- **Advances in Pest Management in Commercial Flowers**
  Editors: Suprakash Pal, PhD, and Akshay Kumar Chakravarthy, PhD
- **Diseases of Fruits and Vegetable Crops: Recent Management Approaches**
  Editors: Gireesh Chand, PhD, Md. Nadeem Akhtar, and Santosh Kumar
- **Management of Insect Pests in Vegetable Crops: Concepts and Approaches**
  Editors: Ramanuj Vishwakarma, PhD, and Ranjeet Kumar, PhD
- **Temperate Fruits: Production, Processing, and Marketing**
  Editors: Debashis Mandal, PhD, Ursula Wermund, PhD, Lop Phavaphutanon, PhD, and Regina Cronje
- **Diseases of Horticultural Crops: Diagnosis and Management, Volume 1: Fruit Crops**

  Editors: J. N. Srivastava, PhD, and A. K. Singh, PhD
- **Diseases of Horticultural Crops: Diagnosis and Management, Volume 2: Vegetable Crops**

Editors: J. N. Srivastava, PhD, and A. K. Singh, PhD

- **Diseases of of Horticultural Crops: Diagnosis and Management, Volume 3: Ornamental Plants and Spice Crops**

  Editors: J. N. Srivastava, PhD, and A. K. Singh, PhD

- **Diseases of Horticultural Crops: Diagnosis and Management, Volume 4: Important Plantation Crops, Medicinal Crops, and Mushrooms**

  Editors: J. N. Srivastava, PhD, and A. K. Singh, PhD

- **Biotic Stress Management in Tomato**
  Editors: Shashank Shekhar Solankey, PhD, and Md. Shamim, PhD

- **Medicinal Plants: Bioprospecting and Pharmacognosy**
  Editors: Amit Baran Sharangi, PhD, and K. V. Peter, PhD

- **Tropical and Subtropical Fruit Crops: Production, Processing, and Marketing**
  Editors: Debashis Mandal, PhD, Ursula Wermund, PhD,
  Lop Phavaphutanon, PhD, and Regina Cronje

ABOUT THE EDITORS

J. N. Srivastava, PhD, has about 20 years of experience of teaching undergraduate, postgraduate, and PhD classes in the courses of plant pathology in the capacity of Assistant Professor/ Junior Scientist (Plant Pathology) at Sher-e-Kashmir University of Agriculture Sciences and Technology, Jammu (J&K), and as Associate Professor cum Senior Scientist (Plant Pathology) at Bihar Agricultural University, Sabour, Bhagalpur, Bihar. His vast experience in agriculture also includes experience in extension programs, consultancy, development, administration, etc.

Dr. Srivastava has extensively pursued research work on different aspects of plant diseases. Dr. Srivastava has made significant research contributions in biological control of diseases in crop plants. He has identified bio-control agents (*Trichodema viride*) to manage guava wilt disease in guava, and he has also done research on integrated disease management of vegetable crops. He is working on biological control/integrated disease management as an eminent scientist and has handled 10 projects as principal investigator and co-principal investigator. In addition to teaching and research, Dr. Srivastava is credited with many publications, including two books, over 45 research papers, over 47 book chapters, over 63 extension articles, four practical bulletins, five technical bulletins, and many leaflets also.

Dr. Srivastava was the recipient of many awards, including Outstanding Achievement Award (2015), Kunwar Sexena Bahadur–SRDA Award (2015), Excellence in Teaching Award (2016, 2017), Dr. R. S. Paroda Medal (2016), Outstanding Scientist Award (2016), Excellence in Science Communication Award (2017), Distinguished Faculty Award (2017, 2018), Eminent Scientist Award (2018), Best Faculty Award (2019), Distinguished Teacher Award (2020). He is also the recipient of several fellowships, including F.H.A.S.-2010, F.B.R.S.-2013, F.S.S.D.A.T.-2014, F.P.P.S.-2015, F.S.E.R.S.-2016, F.S.B.S.R.D.-2016, F.S.A.I.D.-2017, F.P.S.I.-2018, F.N.S.F.-2018, F.I.N.S.O.P.P.,-2019, from various national and international academic/scientific societies.

Dr. Srivastava is a life member of 16 academic/scientific societies, namely, Indian Phyto-pathological Society, Society of Mycology and Plant Pathology, Indian Society of Plant Pathologist, Association of Plant Pathologists of India, Society of Plant Protection Sciences, Society for Recent Development in Agriculture, Society for Scientific Development in Agriculture and Technology, National Academy of Biological Sciences (NABS), Association for Advancement in Plant Protection, Society for Plant Research, Indian Botanical Society, and Society of Human Resource and Innovation. He is also associated with many international/ national/provincial scientific/ cultural/academic/educational bodies.

Dr. Srivastava has served as a member of editorial boards of the *International Journal of Plant Protection.* He is reviewer of various international journals and has also visited Kasetsart University, Thailand. Dr. Srivastava has also attended many national and international workshops. He is also engaged in various agriculture extension activity, namely, farm advisory services, Kisan Chaupal Programme, On Farm Trial (OFT) and Front Line Demonstration (FLD), etc. He has also been involved with exhibitions, campaigns, farmers' fairs, Kishan Gosthi, etc., video conferencing, radio talks, TV talks and vocational training for rural youth (men and women), skill/entrepreneurship development programs, farmer training, in-service training, and field days on maize, sunflowers, rajmash, mash, moong, and soybeans. He is also engaged in Mera Gaon Mera Gaurav.

Dr. Srivastava received his MSc in Agriculture and PhD in Plant Pathology.

A. K. Singh, PhD, is Assistant Professor in the Division of Plant Pathology, Sher-e-Kashmir University of Agricultural Sciences and Technology, Jammu, J&K, India. He has been engaged for more than 12 years in teaching (both undergraduate and postgraduate levels) and research and also involved in the transfer of technology through different extension activities.

He has published more than 30 research papers in national and international journals of repute, one practical manual, and more than 10 book chapters published in books along with eight popular articles. Dr. Singh has also presented many research papers at international, national, and regional symposiums, and seminars. He is the recipient of several prestigious awards for research contributions. He has also served as a member of editorial board of *Krishi Vikas Patrika*, published by Directorate of Exten-sion, SKUAST of Jammu. He is a life member of more than four societies of plant pathologists in India.

He earned his DPhil from the University of Allahabad, Allahabad, UP, India.

CONTENTS

CONTRIBUTORS

Sudheer Kumar Annepu
ICAR-Directorate of Mushroom Research, Solan, Himachal Pradesh, India

Susanta Banik
Department of Plant Paathology, SASRD, Nagaland University, Medziphema, Nagaland 797106, India

Pezangulie Chakruno
Department of Plant Pathology, SASRD, Nagaland University, Medziphema, Nagaland 797106, India

Upma Dutta
Division of Microbiology, Sher-E-Kashmir University of Agricultural Sciences and Technology of Jammu, Jammu, India

N. M. Gohel
Department of Plant Pathology, B. A. College of Agriculture, Anand Agricultural University, Anand, Gujarat 388110, India

Moni Gupta
Division of Biochemistry, Sher-E-Kashmir University of Agricultural Sciences and Technology of Jammu, Jammu, India

Sachin Gupta
Division of Plant Pathology, Sher-E-Kashmir University of Agricultural Sciences and Technology of Jammu, Jammu, India

K. Jayalakshmi
AINRP (Tobacco), ZAHRS, University of Agricultural and Horticultural sciences, Shivmaogga, Karnataka 577201, India

G. Karthikeyan
Department of Plant Pathology, Tamil Nadu Agriculture University, Coimbatore 641003, Tamil Nadu, India

Sudha Jala Kohli
P. G. Department of Biotechnology, Tilkamanjhi Bhagalpur University, Bhagalpur 812001, Bihar, India

Neeraj Kotwal
Division of Entomology, Sher-e-Kashmir University of Agricultural Sciences and Technology, Jammu, Jammu and Kashmir, India

B. K. Prajapati
Directorate of Research, S. D. Agricultural University, Sardarkrushinagar, Gujarat 385506, India

S. Raghu
Crop Protection Division, National Rice Research Institute, Cuttack, Odisha 753006, India

T. Raguchander
Department of Plant Pathology, Tamil Nadu Agriculture University, Coimbatore 641003, Tamil Nadu, India

L. Rajendran
Department of Plant Pathology, Tamil Nadu Agriculture University, Coimbatore 641003, Tamil Nadu, India

J. Raju
Plant Quarantine station, Ministry of Agriculture and Farmers Welfare, Government of India, Mangalore, Karnataka 575011, India

R. Ramjegathesh
Department of Plant Pathology, Tamil Nadu Agriculture University, Coimbatore 641003, Tamil Nadu, India

Mohsinali M. Saiyad
College of Agriculture, Vaso, Anand Agricultural University, Anand, Gujarat 388110, India

A. K. Singh
Division of Plant Pathology, Sher-e-Kashmir University of Agricultural Sciences and Technology of Kashmir (SKUAST-J), Chatha, Jammu 180009, J & K, India

A. K. Singh
Department of Agronomy, Rajendra Prasad Central Agricultural University, Pusa, Samastipur, Bihar, India

S. R. P. Singh
Department of Horticulture, College of Horticulture, Noorsarai, Nalanda 803113, Bihar, India

Priti S. Sonavane
Central Horticultural Experiment Station, ICAR-IIHR, Chettalli 571248, Kodagu, Karnataka, India

J. N. Srivastava
Department of Plant Pathology, Bihar Agricultural University, Sabour 813210, Bhagalpur, Bihar, India

Kavi Sumi
Department of Plant Pathology, SASRD, Nagaland University, Medziphema, Nagaland 797106, India

Baby Summuna
Division of Plant Pathology, Sher-E-Kashmir University of Agricultural Sciences and Technology of Kashmir, Kashmir, India

Ashok Kumar Thakur
P. G. Department of Biotechnology, Tilkamanjhi Bhagalpur University, Bhagalpur 812001, Bihar, India

ABBREVIATIONS

2-DE	2-dimensional electrophoresis
AMPs	antimicrobial peptide
BSR	basal stem rot
CBD	coffee berry disease
CP	coat protein
CRISPR	clustered regularly interspaced short palindromic repeats
CSSV	cocoa swollen shoot virus
CWDEs	cell wall degrading enzyme
DMI	demethylation inhibitors
EBIs	ergosterol biosyntheis inhibitors
EST	expressed sequence tag
ETI	effector-triggered immunity
ETS	effector-triggered susceptibility
FCA	freely available chlorine
FYM	farm yard manure
IPBS	Inter-primer Binding Site Polymorphism
IRAP	Inter-retrotransposon Amplification Polymorphism
LC–MS-MS	liquid chromatography–tandem mass spectrometry
LY	lethal yellowing
MP	movement protein
PAMP	pathogen-associated molecular pattern
PR	pathogenesis-related
PRR	pattern recognition receptor
RIPs	ribosome inhibiting proteins
SA	salicylic acid
SAR	systemic acquired resistance
snciRNA	small noncoding interfering RNA
TMV	tobacco mosaic virus
ULVC	urdbean leaf crinkle virus
UPS	ubiquitin-protease system
VSD	vascular streak dieback
YMC	yellow mosaic virus

FOREWORD 1

Plantation crops constitute a large group of crops. The major plantation crops include coconut, arecanut, oil palm, cashew, tea, coffee and rubber; the minor plantation crops include cocoa. Their total coverage is comparatively less and they are mostly confined to small holdings. However, they play an important role in view of their export potential as well as domestic requirements and in employment generation and poverty alleviation programs, particularly in the rural sector.

India is the third largest producer of coconut and leads 90 coconut-producing countries of the world. It occupies the number one position in arecanut production. India has been considered as a treasure house of valuable medicinal and aromatic plant species. The Government of India have identified and documented over 9,500 plant species, considering their importance in the pharmaceutical industry. Out of these, about 65 plants have large and consistent demand in world trade. As a result, horticulture is not only an integral part of food and nutritional security, but also an essential ingredient of economic security.

Medicinal plants are considered as a rich resources of ingredients that can be used in drug development and synthesis. In addition, these plants play a critical role in the development of human cultures around the whole world. Moreover, some plants are considered as an important source of nutrition, and as a result of that, these plants are recommended for their therapeutic values. These plants include ginger, green tea, walnuts, and some others plants.

Mushrooms are also great sources of protein, fiber, B vitamins (especially niacin), vitamin C, calcium, minerals, and selenium. They also contain antioxidants that are unique to mushrooms, such as ergothioneine, which according to studies, is a highly powerful antioxidant. Mushroom farming is a traditional business, and it has been growing for 200 years. If a farmer wants to gain more profits by low investment, then mushroom farming is the best business in India. To start this business you to need to invest more money. Mushroom is a perishable product so the requirement of close monitoring and some technical skill for mushroom cultivation and mushroom diseases specially management is necessary.

Plantation crops, medicinal crops, and also mushrooms are influenced by their plant diseases as a constraints. In this book, the authors have given a detailed account of the major diseases of important plantation crops, medicinal crops and mushrooms, and their management. This book is substantiated with many illustrations, which are of excellent quality and in fact it is the highlight of the book.

I am sure the book will be found very useful not only by students studying at undergraduate and postgraduate levels, researchers and teachers of plant pathology but also by planners, administrators and growers and various organizations dealing with the export of product of plantation, medicinal crops and mushrooms.

I give my compliment to the authors, Dr. J. N. Srivastava and Dr. A. K. Singh, for their effort on bringing out this publication and wish them success in this and future endeavors.

— Panjab Singh, DSc

President, National Academy of Agricultural Sciences;
Chancellor, RLB Central Agricultural University, Jhansi (U.P.);
President, FAARD Foundation, Delhi, India

FOREWORD 2

Plantation crops include arecanut, coffee, cocoa, cashew, coconut, oil palm, tea, rubber, etc., and are high value commercial crops of greater economic importance and play a vital role in our Indian economy. India enjoys the pride of leading in the production of plantation crops throughout the world. The major plantation crops are coconut, arecanut, rubber, tea, coffee, oil palm, cashew, and cocoa. Today plantation crops assume greater significance in expanding the export base of many third-world countries.

Medicinal plants form a numerically large group of economically important plants that provide basic raw materials for medicines, perfumes, flavors, and cosmetics. These plants and their products not only serve as a valuable source of income for small holders and entrepreneurs but also help the country to earn valuable foreign exchange by way of export.

Mushroom is a nontraditional horticultural crop having high quality of proteins, high fiber value, vitamins, and minerals. There are 61.16 lakh MT of cultivated mushrooms annually produced worldwide. China, with 1.7 million tones of production, accounts for nearly 70% of the world production. Other major mushroom-producing countries are Poland, France, Italy, Indonesia, and Germany. India, with 1.5% contribution, ranks eighth in global mushroom production.

Plantation crops, medicinal crops, and also mushrooms influence by their plant diseases as a constraints. In this book, the authors have given a detailed account of the major diseases of important Plantation crops, Medicinal crops, and also Mushrooms, and their management. This book is substantiated with many illustrations, which are of excellent quality and in fact it is the highlight of the book *Diseases of Horticultural Crops: Diagnosis and Management, Vol. 4: Important Plantation Crops, Medicinal Crops, and Mushroom.*

I am sure the book will be found very useful not only by the students studying at undergraduate and postgraduate levels, researcher and teachers of plant pathology but also by the planners, administrators and growers, and various organizations dealing with the export of fruits from the country.

I congratulate the authors, Dr. J. N. Srivastava and Dr. A. K. Singh for their effort in bringing out this very useful book.

— C. D. Mayee, PhD, DSc

Adjunct Professor, Indian Agricultural Research Institute, New Delhi, India;
President, ISCI, Mumbai, India

FOREWORD 3

India is leading in the total production of certain plantation crops in the world. These include, tea, cashew, arecanut, coconut and rubber. Plantation crops provide direct and indirect employment to many people. e. g. Tea-20 lakhs people-Cashew-5 lakhs people. The plantation production industry supports many byproduct industries and also many rural industries. e. g. Coconut Fiber (obtained from husk) production in India is about 2.2 lakh tones. But diseases are one of the important constraints for production of above plantation crops all over the world including India.

Medicinal plants play a central role not only as traditional medicine but also as trade commodities, meeting the demand of distant markets. Ironically has very small share (1.6%) of this ever-growing global market. However, diseases are pose a significant challenge for production of medicinal crops all over the world, including India.

Mushrooms are also great sources of protein, fiber, B vitamins (especially niacin), vitamin C, calcium, minerals, and selenium. They also contain antioxidants that are unique to mushrooms, such as ergothioneine, which according to studies is a highly powerful antioxidant. Mushroom is the perishable product so the requirement of close monitoring and some technical skill for mushroom cultivation and mushroom diseases especially management practices.

Hence, understanding the fundamental components of disease (i.e. symptoms, causal organisms, disease cycles and epidemiology) is essential for effective management of diseases and to avert significant losses in the yield of plantation, medicinal crops and mushrooms. It is in this context the book *Diseases of Horticultural Crops: Diagnosis and Management, Vol 4: Plantation, Medicinal Crops, and Mushrooms* contains in-depth descriptions of all important disease of plantation, medicinal crops and mushrooms and will be a valuable addition to the literature on plantation, medicinal crops and mushroom pathology. This is a commendable effort has be made by Dr. J. N. Srivastava and Dr. A. K. Singh, Bihar Agricultural University, Bhagalpur, Bihar, and Sher-e-Kashmir University of Agricultural Sciences and Technology, Jammu, (J&K), in planning, compiling and editing this publication.

I am sure the book will be found very useful not only by the students studying at under graduate and post graduate levels, researcher and teachers of Plant Pathology but also by the planners, administrators and growers and various organizations dealing with the export of Plantation, Medicinal Crops and Mushroom from the States/country.

I gives compliment to the authors Dr. J. N. Srivastava and Dr. A. K. Singh for their effort on bringing out this publication and wish them success in this and future endeavors.

I heartily congratulate the editors/authors Dr. J. N. Srivastava and Dr. A. K. Singh for bringing out this useful publication that provided comprehensive information about Plantation, Medicinal Crops and Mushroom diseases.

— Amar Nath Mukhopahyay, PhD, DSc

Ex-Vice Chancellor, Assam Agricultural University, Jorhat, Assam, India;
Ex-Director General, Tea Research Association India;
Ex-Director, Biopesticide Division, Super Agro India Pvt. Ltd., Kolkata, India

PREFACE

Plantation crops include arecanut, coffee, cocoa, cashew, coconut, oil palm, tea, rubber, etc. are high value commercial crops of greater economic importance and play a vital role in our Indian economy. The main drawback of plantation crops in India is that major portion of the area is of small holdings (except tea), which hinders the adoption of intensive cultivation. Main products and by-products not only have export prospects but also have considerable internal demand in several ancillary industries. Earnings from export of plantation crops account for 27% of total agricultural commodities and 4.8% of total export.

Medicinal and aromatic plants form a numerically large group of economically important plants that provide basic raw materials for medicines, perfumes, flavors, and cosmetics. These plants and their products not only serve as a valuable source of income for small holders and entrepreneurs but also help the country to earn valuable foreign exchange by way of export. Medicinal plants are those plants that are rich in secondary metabolites and are potential source of drugs. These secondary metabolites include alkoloids, glycosides, coumarins, flavonides, steroids, etc. India is one of the few countries where almost all the known medicinal plants can be cultivated in some part of the country or the other.

Mushroom is a nontraditional horticultural crop having high quality proteins, high fiber value, vitamins, and minerals. World produces 61.16 lakh MT of cultivated mushrooms annually. The share in production of different types of mushrooms worldwide is button (31%), shiitake (24%), oyster (14%), black-ear mushroom (9%), paddy straw mushroom (8%), and milky/others (the rest). China produces 63% of the world production of mushrooms and ranks first among world's mushroom-producing countries.

India produces more than 70,000 MT of mushroom. In India button, oyster, milky, and paddy straw mushrooms are grown but button mushroom contributes the highest share of production. Even though button mushroom cultivation started in India in the 1970s at Chail and Kasauli (Himachal Pradesh, India), it is now grown all over the country, mostly in tropical areas, where raw materials and labor are available at competitive rates as compared to hilly regions. The mushrooms produced in the large commercial units in

India are processed and packed in cans/jars for export. This is necessary as button mushroom has short shelf life of less than a day.

In spite of all above facts, the production from some plantation, medicinal crops, and mushroom is less as compared to advanced countries because of different reasons; the most important being the prevalence of diseases in plantation, medicinal crops, and mushroom. The management of any disease successfully involves its detailed study regarding symptoms, casual agent, disease cycle and epidemiology.

This book entitled, *Diseases of Horticultural Crops: Diagnosis and Management, Volume 4: Plantation, Medicinal Crops, and Mushrooms,* includes 13 chapters. Each chapter has a detailed account that includes an introduction, symptoms, causal organisms, disease cycles, epidemiology, and management of economically important diseases.

The book chapters have been contributed by authors who are engaged in teaching, research, and extension services and are well-known national scientists in their respective field. The authors, while writing the chapters, have incorporated their experience and knowledge with the recent development in field of plant diseases.

It is hoped that the book will cater to the needs of students studying at undergraduate and postgraduate levels, researchers, teachers, planners, administrators, and growers not only in the discipline of plant pathology but also in the other fields of agriculture.

We sincerely acknowledge our thanks and gratitude to the esteemed scientists who have spared their time and contributed valuable chapters for this book. We are also heartily thankful to Apple Academic Press for publishing this book.

— Dr. J. N. Srivastava
Dr. A. K. Singh

PART I
Plantation Crops

CHAPTER 1

PRESENT SCENARIO OF DISEASES IN ARECA NUT OR BETEL NUT (*ARECA CATECHU* L.) AND THEIR MANAGEMENT

J. RAJU[1], K. JAYALAKSHMI[2*], PRITI S. SONAVANE[3], and S. RAGHU[4]

[1]*Plant Quarantine station, Ministry of Agriculture and Farmers Welfare, Government of India, Mangalore, Karnataka 575011, India*

[2]*AINRP (Tobacco), ZAHRS, University of Agricultural and Horticultural sciences, Shivmaogga, Karnataka 577201, India*

[3]*Central Horticultural Experiment Station, ICAR-IIHR, Chettalli 571248, Kodagu, Karnataka, India*

[4]*Crop Protection Division, National Rice Research Institute, Cuttack, Odisha 753006, India*

**Corresponding author. E-mail: jayalakshmipat@gmail.com.*

ABSTRACT

Areca nut palm (*Areca catechu* L.) belongs to the family *Arecaceae* which grows in much of the tropical Pacific, Asia, and parts of East Africa. Palms have large, evergreen leaves that are either fan or feather-leaved compound and spirally arranged at the top of the stem. It is a common source of chewing nut, known as betel nut or *supari*. India is the largest producer and consumer of areca nut, and it is extensively used by many people linked with religious practices. Karnataka (40%), Kerala (25%), Assam (20%), Tamil Nadu, Meghalaya, and West Bengal are the major areca-nut-producing states of India. Areca nut is infected with the various diseases which are mentioned in the various sections of this chapter.

1.1 FUNGAL DISEASES

1.1.1 LEAF BLIGHT/GRAY BLIGHT

Symptoms:

1. During summer months, seedlings and young palms will be affected (Roy, 1965).
2. Affected leaves show reddish brown or dark-brown-discolored spots on mature outer whorls (Hossain et al., 1992).
3. Later spots coalesce to form irregular patches with brown margin and gray center with minute black dot-like structures (Aderungboye, 1974).
4. At the advanced stage, extensive blighting of leaves occurs (Agnihotri, 1963).

FIGURE 1.1 Gray blight.

Causal Organism: *Pestalotia palmaraum, Phomopsis palmicola*

Disease Cycle:

1. Primary source: Affected plant debris
2. Secondary source: Airborne Conidia

Epidemiology:

During monsoon, high humidity and low temperatures (20 °C–25 °C) favor the disease development (Bhat et al., 1992b).

Integrated Disease Management:

1. Removal and destruction of affected older leaves.
2. Application of recommended fertilizers.
3. Maintenance of proper drainage in a garden.
4. Providing shade to nursery plants.
5. Spray 1% Bordeaux mixture or 0.4% Mancozeb (4 g/L).

1.1.2 LEAF SPOT DISEASE

Symptoms:

1. All ages of palms are affected.
2. Initially, small yellow specks appear on the leaf lamina; later, specks enlarge to form irregular sunken lesions with dark brown margins and light brown center having minute dot-like structures.
3. Around the sunken lesions, a prominent yellow halo develops. As infection progresses, spots coalesce to form large patches, leading to premature yellowing, blighting, drying, drooping, and shredding of leaves (Saraswathy et al.,1977).
4. In severe condition, stunted growth and death of the seedlings have been observed.

Causal Organism: *Colletotrichum gloeosporioides.*
Acervuli asexual fruiting structures develop on older sunken spots and appear as brown to black dots. The conidia are straight, cylindrical or oval, hyaline with two oil drops, and are nonseptate with round ends which borne on hyaline conidiophores. The perithecial stage of the fungus is not very common (Chowdappa et al., 1995).

Disease Cycle:

1. Primary source: Inoculum remains on dried leaves, defoliated branches.
2. Secondary source: Secondary spread is through airborne conidia.

Epidemiology:

Pathogen requires infection optimum temperature of 25 °C for infection and relative humidity of 95%–97% (Alice et al., 1985).

Integrated Disease Management:

1. Removal and destruction of affected plant debris.
2. Providing shade to nursery plants.

3. 2–3 times spraying 1% Bordeaux mixture or Mancozeb 75 WP at 2 g/L or copper oxycholride 50 WP 3 g/L in 15-day intervals (Narasimhan, 1924).
4. In severe cases, 2–3 sprays of hexaconazole or propiconazole at 1 mL/L in15-day interval provide better management.

1.1.3 STEM BLEEDING DISEASE

Symptoms:

1. 10–15-year-old palms are more susceptible.
2. Initially, small discolored depressions or spots near the stem basal portion.
3. Spots coalesce to form larger patches followed by cracking.
4. Fibrous layers of the stem disintegrate which eventually hollows up to varying depths along the infected portions (Bavappa and Sahadevan, 1952).
5. Dark brown gummy exudates ooze out from cracks.
6. Stem tapers near the crown.
7. Crown size and yield are reduced.
8. Death of the palms is observed in a severe condition.

FIGURE 1.2 Stem bleeding.

FIGURE 1.3 Stem breaking.

Causal Organism: *Ceratocystis paradoxa* (*Thielaviopsis paradoxa*).

Disease Cycle:

1. Primary source: Affected cracks and cavities of infected stem carrying perithecia and chlamydospores.
2. Secondary source: Conidia dispersed through irrigation water, insects, wind.

Epidemiology:

1. Heavy rainfall followed by drought leads more susceptibility.
2. Presence of cracks and insect wounds (*Diocalandra* and *Xyleborus* beetles) on stem leads severe infection.
3. Physiological imbalances, poor fertilization, and poor maintenance of orchard.

Integrated Disease Management:

1. Avoiding wounds or mechanical injury to the stem.
2. Providing proper drainage facilities to the garden during rainy season.
3. Trailing pepper on the palms.

4. Application of 50 kg of organic manure, 5 kg neem cake with *Trichoderma* to the basin during September. Scraping of the affected tissues and dress with hot coal tar or Bordeaux paste (Chandramohanan and Kaverappa, 1985).
5. In older trees, after draining the fluid, cavities filled with mixture of tar coal and sawdust or bordeaux paste to prevent the disease.
6. Application of Carbaryl 50% WP on the trunk at 3 g/L to manage stem boring insects (Ramakrishnan, 1990).
7. Swab Hexaconazole after the removal of the infected portion along with healthy tissues and drench soil with 0.1% Hexaconazole or root feeding of 100 mL of Hexaconazole 2.0% through 2–3 roots (Chandramohanan, 1979).

1.1.4 INFLORESCENCE DIE BACK/BUTTON SHEDDING DISEASE

Dieback of inflorescence and button shedding is one of the reasons for a low fruit set in areca palms. The disease is seen throughout the year but becomes severe during summer months, that is, from February to May. Yellowing, browning, and shedding of buttons are accelerated by water stress and high temperature that prevail during summer.

Symptoms:

1. Yellowing of rachis of inflorescence.
2. Yellowing progresses from the tip of the rachis toward the main rachis.
3. As the disease progresses, the entire rachis turns dark brown causing wilting and drying of inflorescence (inflorescence die back).
4. Subsequent spread of yellowing and discoloration induces the shedding of female flowers (buttons) in large numbers.
5. Discoloration of the inner soft calyx region of the female flowers and buttons results in their shriveling, drying, and shedding.
6. Direct infections on the stigmatic end of the female flowers causes discoloration and shedding without showing any typical symptoms mentioned above.
7. Concentric rings of light-pink-colored mass of spores appear on the discolored infected inflorescence (Bopaiah and Koti Reddy, 1982).
8. On incubation, ash-colored mycelial growth and pink-colored spores of fungus develop from the fallen button and discolored rachis/ inflorescence.

9. On severe infection, the female flowers and buttons shed completely or dry up and remain on the crown.

FIGURE 1.4 Inflorescence dieback.

Causal Organism: *Colletotrichum gloeosporioides*.

Disease Cycle:

Debris of previous years' infected bunches for eight months acts as a primary source of inoculum. Secondary spread is through wind-borne conidia.

Epidemiology:

The optimum temperature of 25 °C and relative humidity of 95%–97% favor the disease infection and development.

Integrated Disease Management:

1. Removal and destruction of affected and dried inflorescence, buttons, and all fallen nuts.
2. Spraying of Zineb or Mancozeb (4 g/L) twice or 1% Bordeaux mixture and Captan (3 g/L), just after the setting of female flowers and then again at the interval of 3–4 weeks as prophylactic measure.
3. Application of 0.1% hexaconazole or propiconazole or 50 ppm (50 mg/L) of aureofungin or 0.25% Antracol.

1.1.5 BASAL STEM ROT/FOOT ROT DISEASE

Symptoms:

1. Disease starts with yellowing or pale discoloration and drooping of the outer whorl of leaves, followed by the exudation of reddish-brown liquid through cracks at the base of the trunk and oozing spread upward.
2. Later, the inner whorl of leaves also becomes yellow, leaving only the spear leaf green which results in aborting the development of inflorescence and nuts (Sarma and Murthy, 1971).
3. In severe conditions, the spindle gets dried up and the crown topples down, leaving the bare stem or tapering of stem. Reduction in inter-nodal length leads to complete damage of internal tissues (Banerjee and Sarkar 1958).
4. Further, the infected brittle stem breaks off easily during heavy wind.
5. Decaying of tissues at the bleeding point and rotting of the basal portion of the stem occurs. The bark turns brittle and often gets peeled off in flakes, leaving open cracks and crevices. The internal tissues are discolored and disintegrated which become brittle and dry with a musty smell. Uptake of water and nutrients is affected (Banerjee and Sarkar 1959).
6. Bracket formation at the base of the trunk during rainy season. Ultimately, the palm dies off.

Causal Organism: *Ganoderma lucidum, G. applanatum*

Pathogen produces hyaline septate mycelium with clamp connections. Either terminal or intercalary, it produces chlamydospores. The fruiting body is perennial, usually lateral, and is corky first, and later it becomes woody. Basidiospores are brown, thick walled, and truncated at one end (Sampath Kumar and Nambiar, 1990).

FIGURE 1.5 Basal stem rot.

Disease Cycle:

1. Primary source: Infected stumps, brackets containing basidiospores, pathogen having wide host range, namely mango, jack, citrus, coconut, coffee, and tea (Sampath Kumar and Nambiar, 1993).
2. Secondary source: Irrigation water and by root contact (Rhizomorph) (Sampath Kumar and Nambiar, 1996a).

Epidemiology:

The disease is highly prevalent in sandy soils and where areca nut gardens are raised under rainfed conditions. The lack of soil moisture during summer months, the presence of old infected stumps in the garden, injury to roots, and nonadoption of recommended cultural practices favor the disease infection (Bose, 1931; Koti Reddy and Saraswathy, 1976).

Water stress condition in soil during summer and water stagnation condition during rainy season favor the spread of the disease. This disease is observed in 20–30-year-old palms (Butler, 1906).

Integrated Disease Management:

1. Removal and destruction of severely affected parts and stumps of dead palms, including roots.

2. Drench around the base of the infected palm with 0.1% hexaconazole (Sampath Kumar and Nambiar, 1996b).
3. Provide proper drainage facilities in the garden.
4. Avoid dense planting.
5. Avoid flood irrigation.
6. Avoid repeated ploughing and digging in the affected gardens.
7. Balanced application of manures and fertilizers.
8. Application of 2–3 kg neem cake/palm/year.
9. Avoid growing of collateral hosts, namely *Delonix regia* and *Pongamia glabra* in the vicinity of gardens (Lalithakumari, 1969).
10. Soil application of *Trichoderma harzianum* with farmyard manure as prophylactic measure.
11. Root feeding of palms with 0.1% hexaconazole to 2–3 roots in the early stage of infection at three months' intervals (Koti Reddy, et al., 1978).
12. Digging trenches 60 cm deep and 30 cm wide around the "Anabe"-infested areca palms and drenching with 0.3% Captan at 2 g/L of water prevent the spread of the disease (Nair and Rao, 1965; Vinayaka and Gunasekaran, 2004).

1.1.6 KOLEROGA OR MAHALI OR FRUIT ROT DISEASE

Symptoms:

- **Neergole:** Early-stage symptom
- **Boosargole:** Later-stage symptom

1. Initially, dark green/yellowish water-soaked lesions appear on the surface of nuts near the perianth (calyx), and infected nuts lose their natural green luster (Anandaraj, 1985).
2. Characteristic symptom is rotting and extensive shedding of the immature nuts which lie scattered near the base of the tree.
3. The lesions on nuts gradually spread covering the entire nuts (before or after shedding) which consequently rot and shed.
4. White mycelial mass envelopes on the entire surface of the fallen nuts. As the disease advances, the fruit stalks and the axes of the inflorescence rot and dry, sometimes being covered with white mycelial mats (Coleman, 1911).
5. Affected nuts are lighter in weight having large vacuoles. The dark brown radial strands appear on kernel, making them unfit for chewing.

6. Dry Mahali symptoms that appear later in the season cause rotting and drying of nuts without shedding.

FIGURE 1.6 Fruit rot.

Causal Organism: *Phytophthora meadii, P. arecae*

Disease Cycle:

Primary source is the affected host debris or fallen nuts carrying dormant mycelium, oospores over summer. With the onset of monsoon rains, the fungus becomes active and produces cottony mycelium that infects the tender host tissue. Heavy wind and, to a certain extent, small insects and rain splashes spread the inoculum. Under favorable conditions, zoospores released from the sporangia germinate in the films of water and penetrate the nut surface through the stomata (Bharati and Anahosur, 1994a, 1994b).

Epidemiology:

Southwest monsoon plays a key role in occurrence, survival, and spread of the disease. Continuous heavy rainfall with intermittent bright sunshine hours, 20 °C–23 °C temperature, and >90% relative humidity is congenial for disease development. The affected fruits bunch toward the end of the rainy season may remain mummified on the palm, and such nuts provide inoculums for bud rot or crown rot disease or the recurrence of fruit rot in the next season (Dutta and Hegde, 1987).

Integrated Disease Management:

1. Before the onset of monsoon, spray 1% Bordeaux mixture on the bunches as a prophylactic measure. After spraying, cover the areca bunches with polythene covers (125–200 gauge 24 × 30 inches) (Narasimhan, 1924).
2. The second spray of 1% Bordeaux mixture or copper oxychloride of 3 g/L should be done after an interval of 40–45 days. If monsoon prolongs, third spray should be applied (Anandaraj and Saraswathy, 1986).
3. A fine spray is needed for effective spread of 1% Bordeaux mixture over the surface of the nuts. Spraying operations are to be undertaken on clear sunny days (Narasimhan, 1928).
4. Collect and destruct all fallen and infected nuts.
5. Severe incidence of fruit rot during monsoon may lead to the incidence of bud rot and crown rot diseases. Hence, preventive measures to be taken up to control these diseases as well.
6. Cut the affected region from the crown and treat the wound/cut end with 10% Bordeaux paste.
7. Cover the treated bud with protective covering till the normal shoot emerges (Chowdappa et al., 2000b).

1.1.7 BUD ROT/CROWN ROT DISEASE

It is a fatal disease and appears during heavy rainfall and in mydan areas during rainy and winter seasons (Iswara Bhat, 1965).

Symptoms:

1. The initial symptom is the discoloration of spindle leaf green color to yellow and then to brown, followed by withering.
2. The leaves rot, and the growing bud rots cause the death of the palm.
3. The affected young leaf whorl can be easily pulled off.
4. The outer leaves also become yellow and droop off one by one, leaving a bare stem.
5. Infection spreads inside the bud causing the rotting of the growing bud, and the surrounding leaves result in the death of the palm.
6. The affected young leaf whorl can be easily pulled off.

7. Infection spreads to the adjacent or outer leaves become yellow and droop off one by one, leaving a bare stem.
8. A secondary organism colonizes the infected tissues, converts it into a slimy mass, and emits foul smell.
9. Severe fruit rot infection spreads to the stem and outer leaf sheaths. An inner portion of the leaf sheath exhibits water-soaked lesions, and the base of leaf sheath rots results in yellowing of outer leaves.

FIGURE 1.7 Bud rot.

Causal Organism: *Phytophthora meadii, P. arecae* (Butler and Bisby, 1931).

Disease Cycle:

The pathogen survives in the form of dormant mycelium on the treetop either on the infected, dried bunches or bud-rot-affected dead palms or upper layers of soil or in the form of oospores on fruit-rot-affected nuts under the natural condition. These are the primary sources of inoculum. Secondary spread takes place through zoospores by rainwater or wind and insects.

Epidemiology:

South west monsoon plays a key role in occurrence, persistence, and spread of fruit rot. Continuous heavy rainfall with intermittent bright sunshine hours, low temperature (20 °C–23 °C) and high relative humidity (>90%) are factors congenial for disease development.

Integrated Disease Management:

1. Removal and destruction of dead palms or crowns affected by bud/ crown rot and fruit-rot-affected dried bunches and fallen nuts.
2. Apply Bordeaux paste to the bud region and cover with a polythene cover or a mud pot.
3. Spray and drench 1% Bordeaux mixture to spindle the leaf base (Bakshi, 1974).
4. Spray Mancozeb 75 WP 3 g/L or Zineb 4 g/L on the opening of female flowers in most of the inflorescences. This should be followed by a second spray after 25 days (Lingaraj, 1969).
5. In the crown or above bud region, keep a mixture of $CuSO_4$ + NaCl (1:3) in a cloth bag.

1.2 BACTERIAL DISEASES

1.2.1 BACTERIAL LEAF BLIGHT

The disease occurs in an endemic form in the maidan parts of Karnataka in 3–5-year-old plants (Coleman and Rao, 1918).

Symptoms:

Initially, on leaves, dark green water-soaked lesions appear, translucent linear lesions or strips on the leaf lamina. On the corresponding lower surface, the lesions show a creamy white exudate. The wet exudate is slimy, but on drying, it becomes waxy and leads to necrosis of the leaves. The infection may cause partial or complete blighting of the leaf, and in extreme cases, the plants succumb to infection (Patel and Rao, 1958).

Causal Organism: *Xanthomonas arecae*

Disease Cycle:

The bacterium does not remain in the soil for more than 75 days, indicating that soil is not the primary source of inoculum (Sampath Kumar, 1993). The bacterium survives on the blighted leaves and leaflets remaining on the palm. The pathogen was found to confine to the green or yellowish areas adjacent to the blighted portions and remain viable in such conditions during the off season, thus acting as a source of inoculum for epidemics during conducive periods (Rao and Mohan, 1976).

Epidemiology:

The disease is very severe during rainy months, that is, from July to October, when the average monthly rainfall is 130 mm or above with more than 10 rainy days/month. Temperature above 30 °C and below 17 °C will slacken the disease spread (Sampath Kumar, 1985).

Integrated Disease Management:

1. Remove and destroy the affected leaves.
2. Spray 3 g copper oxychloride and streptocycline 500 ppm/L as prophylactic spray.

1.3 MYCOPLASMAL DISEASES

1.3.1 YELLOW LEAF DISEASE

Symptoms:

1. The symptoms of the disease appear as translucent spots and characteristic yellowing of the tips of the leaflets in two or three leaves of outermost whorls (Ponnamma and Solomon, 2000).
2. Yellowing gradually extends to the middle of the lamina showing a clear-cut demarcation of yellow and green parallel bands on both sides of the midrib of leaflets (Mathai, 1976).
3. The affected leaves often develop necrosis from their tips.
4. The diseased leaves possess smaller epidermal cells, stomatal pores, and midrib parenchyma cells. Degeneration of cortex and presence of tyloses are commonly seen in the diseased roots (Nayar, 1976).
5. In advanced stages, the leaves are reduced in size and become stiff and pointed. The crown gets reduced and ultimately falls off. Tips

and absorbing regions of young roots become dark and gradually rot (Radhakrishnan Nair, 1994). The kernel of affected nuts shows discoloration and later turns blackish. Such nuts are unsuitable for chewing (Chowdappa et al., 2000a).

6. Finally, the crown leaves fall off, leaving a bare trunk. Tips of absorbing young roots blacken and rot, especially in water-logged conditions. Roots exhibit vascular discoloration (Chowdappa et al., 1995).

FIGURE 1.8 Yellow leaf disease.

Causal Organism: *Phytoplasma*

Transmission occurs through an insect vector: Plant hopper (*Proutista moesta*)

Integrated Disease Management:

1. Application of the recommended dose of fertilizers at 100:40:140 g/palm/year in two equal splits during September and October and February and March (Guruswamy and Krishnamurthy, 1994).
2. Add 1 kg of rock phosphate along with 2 kg of neem cake/palm at two intervals into infected palms to increase new root formation.
3. During April, polythene cover containing 2 g phorate granules should be tied to the palm for managing the spindle bug.
4. Soil application of lime at 500 g/palm once in 2–3 years during June and July.
5. Application of 12 kg of organic manure, compost, and green leaves per palm per year.

6. Provide proper irrigation during summer months and avoid water stagnation in the garden by providing drainage facilities.
7. Growing of cover crops in the garden.
8. Remove the affected palms and replant with high-yielding local varieties such as Kasargod local, Dakshin Kannada local, and Sirsi local; and also with dwarf varieties such as Hirahalli dwarf and Mohit nagar.
9. Spray dimethoate 30 EC (1.5 mL/L) to manage the plant hopper.

1.4 NEMATODES DISEASES

1.4.1 THE BURROWING NEMATODE DISEASE

Burrowing nematode is the most important plant parasitic nematode having a wide host range, namely coconut, areca nut, black pepper, banana, and betel vine (Patel and Rao, 1958). This nematode is widely distributed in South India and is also reported from Lakshadweep Island. The nematode has a very wide host range covering more than 300 species of plants including trees, shrubs, and herbs. Ginger, turmeric, sweet potato, sugarcane, groundnut, avocado, nutmeg, tea, coffee, and cardamom are other important crops known to be susceptible to this nematode. Burrowing nematode is a notorious pathogen that causes the toppling disease of banana in all banana growing regions. Parasitization by *Radopholus similis* (*R. similis*) causes gross reduction in the quality and quantity of the yield.

Symptoms:

1. Infested plants show yellowing of the leaves and stunted growth of the plants.
2. These above-ground symptoms are nonspecific, and the only definite method to identify an infested plant is to look for symptoms on fresh roots.
3. The nematodes produce small, elongate, orange-colored lesions on the creamy white to light-orange-colored portion of the main tender roots, and these lesions coalesce and cause extensive root rotting.
4. Tender roots on heavy infestation become spongy in texture.
5. As high as 4000 nematodes are harbored by 1 g (1″ length) of main areca nut roots. Lesions and rotting are more on the tender portions of roots and do not normally occur on the hard, older red, or brown-colored portion of roots.

Causal Organism: *R. similis.*

Biology:

The nematode takes three weeks to complete its life cycle from an egg to an adult at a temperature range of 24 °C–32 °C. All juveniles and females are infective except males. They enter into the tender roots and feed in the cortical region of coconut and areca nut roots. It is a migratory endoparasite and is disseminated mainly through the infested planting materials, floods, irrigation water, farm implements, and bulk transport of soil (Coleman, 1910).

Epidemiology:

Being moisture dependent, the population fluctuations of nematodes are markedly affected during summer months. Maximum nematode populations occur from September to November and minimum from April to June. The nematode is known to survive and increase best in deep, well-drained sandy loam soils but less in shallow, poorly drained clayey soils.

Integrated Disease Management:

1. Use nematode-free planting materials or disease-free seedlings.
2. Raise seedlings in polybags, containing fumigated or sterilized soil (Daji, 1948).
3. Burn planting pits, avoiding running water from infested fields to noninfested fields, removing soil from agricultural implements, while transporting from a nematode-infested fields to noninfested field is helpful in preventing spread of the nematodes.
4. Apply neem cake of 1.5 kg/palm to reduce the nematode populations.
5. Apply carbofuran 3G of 10 g/palm to reduce the nematode populations.

1.5 PHYSIOLOGICAL DISORDERS

1.5.1 CROWN CHOKE/BAND/HIDIMUNDIGE DISORDER

Symptoms:

1. It is a physiological disorder that occurs due to adverse environmental conditions, resulting in the reduction of leaf size which turns brittle and crinkled with wavy margins (Dorasami, 1956).

2. As the disorder advances, there is a reduction in internodal length, formation of small bunches, and tapering of stems.
3. In severe cases observed, crown bends and forms the rosette shape due to the failure of natural opening of leaves.
4. In the affected palms, bunches become small and malformed, and the roots are poorly developed, crinkled, and brittle (Raghavan and Baruah, 1956).

FIGURE 1.9 Crown choke.

Cause:

Poor drainage, low soil fertility, or adverse environmental conditions.

Management:

1. Provide good drainage and adequate aeration to the roots by removing hard soil pan.
2. Management of spindle bug, mealy bugs, scales, and mites reduces the disorder.

3. At the base of affected palms, apply powder formulation of copper sulphate and lime in equal quantities of 225 g/palm twice a year (Rao, 1960).
4. Apply borax of 25 g/palm to manage the disorder.

1.6 NUT-SPLITTING DISORDER

Symptoms:

1. Palms in the age group of 10–25 years are more susceptible.
2. Initial symptom appears as premature yellowing of nuts when they are half to three-fourth mature (Ramakrishnan, 1956).
3. This is followed by splitting of nuts from both sides and the tips which expand longitudinally toward the calyx exposing the kernel (Saidalikutty, 1951).

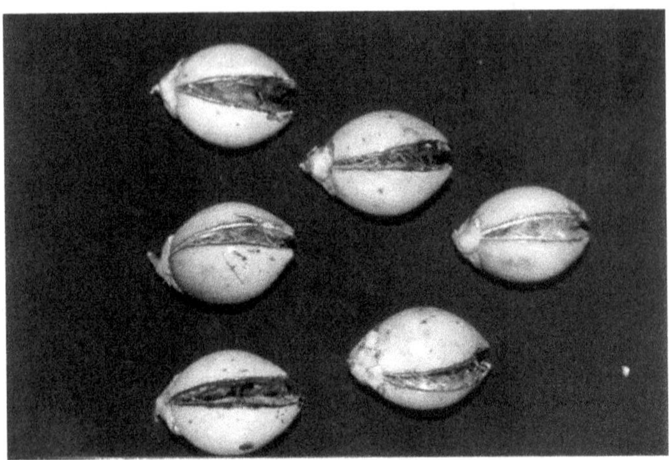

FIGURE 1.10 Nut splitting.

Cause:

This is a physiological disorder. Sudden flush of water after a period of water stress is the main cause. Boron deficiency also causes this disorder (Iswara Bhat, 1961).

Management:

1. Improvement of drainage and spray of Borax of 2 g/L in the initial stages of disorder are found effective in controlling the disease.

1.7 SUN SCORCH AND STEM BREAKING

Symptoms:

1. It is a disorder, appear from the prolonged exposure of palms to severe sunlight. Palms exposed to the southwest sun are more prone to stem breaking.
2. Initially, the symptoms appear as golden yellow splits on the exposed side of the stem which turn dark brown. Further, colonization by saprophytic fungi weakens the stem and finally breaks.

Cause:

Prolonged exposure of palms to severe sun light or high temperature.

Management:

1. Rising of tall, fast-growing varieties on southwest sides of the garden.
2. Wrapping stems with areca sheath/leaves to protect the palms from the southwest sun
3. Apply whitewash to the exposed stem portion.
4. Trail pepper vines on the stem of the palms to avoid direct sunlight on the stem.

KEYWORDS

- **Areca nut**
- **etiology**
- **symptomatology**
- **management**

REFERENCES

Aderungboye, E. (1974). Occurrence and severity of the blast disease on exotic palms in Nigeria. *J. Nigerian Inst. Oil. Palm Res.* 5: 73-77.

Agnihotri, V. P. (1963). Some pathological studies of Alternaria tenuis causing leaf spot disease of Areca catechu. *Mycopathol. Mycol. App.* 21: 64–73.

Alice, K. J., Balakrishnan, S. and Karunkaran, P. (1985). A comparative study of microflora of healthy and diseased arecanut palms affected by yellow leaf disease. In: *Arecanut Research and Development* (Eds. Shama Bhat, K. and Radhakrishnan Nair, C. P.). Kasaragod: CPCRI, pp. 107–113.

Anandaraj, M. (1985). On the mechanism of spore dispersal in Phytophthora arecae the causal organism of "Koleroga" of arecanut. In: *Arecanut Research and Development* (Eds. Shama Bhat, K. and Radhakrishnan Nair, C. P.). Kasaragod: CPCRI, pp. 83–85.

Anandaraj, M. and Saraswathy, N. (1986). Effect of systemic fungicides on fruit rot of arecanut. *Indian Phytopathol.* 39: 607–609.

Bakshi, B. K. (1974). Control of rot disease in reforested stands with special reference to Khair, Sisso, Eucalyptus etc. *Indian Forester.* 100: 71–80.

Banerjee, S. and Sarkar, A. (1958). Studies on heterothallism IV. *Ganoderma lucidum* (Leys) Karst. *Sci. & Cult.* 24: 193–195.

Banerjee, S. and Sarkar, A. (1959). Spore forms in sporophores of *Ganoderma lucidum* (Leys) Karst. *Pm. Indicin. Acad. Sci.* B. 49: 95–98.

Bavappa, K. V. A. and Sahadevan, I. C. (1952). A note on fruit cracking in arecanut. *ICAC Monthly Bull.* 3: 20–21.

Bharati, H. and Anahosur, K. H. (1994a). Studies on variability in isolates of *Phyllostictu arecae Hohnel* (morphology and pathogenicity). hataka. *J. Agric. Sci.* 7: 181–185.

Bharati, H. and Anahosur, K. H. (1994b). Studies on variability in isolates of *Phyllosticta arecae Hohnel* XI. Physiological and nutritional studies. hataka. *J. Agric. Sci.* 7: 431–436.

Bhat, R., Hiremath, P. C. and Hegde, R. K. (1992b). A note on the incidence and severity of leaf blight of arecanut caused by *Phyllosticta* arecae in Karnataka. Karnataka. *J. Agric. Sci.* 5: 402–403.

Bopaiah, B. M. and Koti Reddy, M. (1982). Distribution of microflora population in the rhizosphere of arecanut. *J. Plantn. Crops*. 10: 127–128.

Bose, S. R. (1931). Short note on the abnormal anatomy of sporophores of *G. lucidum*. In: *Proceedings of Eighteenth Indian Science Congress Association*, Calcutta. pp. 263.

Butler, E. J. (1906). Some diseases of palm. *Agric. J. India*. 1: 299–310.

Butler, E. J. and Bisby, G. R. (1931). Fungi of India. *Science Monograph*. Indian Council of Agricultural Research. pp. 273.

Chandramohanan, R. (1979). Effect of soil application of seven chemicals on disease incidence and high yield of YLD affected areca palms. In: *Proc. PLACROSYM II*. CPCRI, Kasaragod. pp. 361–366.

Chandramohanan, R. and Kaverappa K. M. (1985). Epidemiological studies on die-back f Arecanut caused by Colletotrichum gloeosporiodes. In: *Arecunut Research and Development* (Eds. Shama Bhat, K. and Radhakrishnan Nair, C. P.). Kasaragod: CPCRI, pp. 116–119.

Chowdappa, P., Balasimha. D., Rajagopal, K. and Ravindran, F. S. S. (1995). Stornatal response of arecanut palms affected with yellow leaf disease. *J. Plantn. Crops.* 23: 116–121.

Chowdappa, P., Daniel, E. V., Balasimha, D. and Mathai, C. K. (2000a). "Physiology" in a Arecanut Yellow Leaf Diseases. (Eds. Narnpoothiri, K. U. K., Ponnarnma, K. N. and Chowdappa, P.). *Technical Bulletin*, No. 39, Kasaragod: CPCRI, pp. 24- 37.

Chowdappa, P., Saraswathy, N., Vinayagopal, K. and Somala, M. (2000b). Control of fruit rot of arecanut through polythene covering. *Indian Phytopathol*. 53: 321.

Coleman, L. C. and Rao, M. K. V. (1918). *The Cultivation of Areca Palm in Mysore*. Bangalore: Department of Agriculture, Mysore State, p. 32.

Coleman, L. C. (1910). Diseases of the areca palm. 1. "*Kole roga*", Mycological Series. *Bulletin*, No. 2. Bangalore: Department of Agriculture, Mysore State, p. 92.

Coleman, L. C. (1911). "*Anabe roga*" of "supari". *Annual Report for 1909–1910*. Bangalore: Agricultural Chemist Mysore, Department of Agriculture, p. 32.

Daji, J. A. (1948). Manganese toxicity as a probable cause of the band disease of areca palms. *Curn. Sci.* 17: 259-260.

Dorasami, L. S. (1956). A brief note on arecanut industry in Mysore State. *Arecanut J.* 7: 58–61.

Dutta, P. K. and Hegde, R. K. (1987). Studies on two Phytophthora diseases (Koleroga of arecanut and black pepper wilt) in Shimoga district, Karnataka State. *Plant Pathol. Newsl.* 5: 1–2, 29.

Guruswamy, K. T. and Krishnamurthy, N. (1994). Characterization of soils collected from profiles in yellow leaf disease affected arecanut gardens in Thirthahalli Taluk (Karnataka, India). *J. Agric. Sci.* 7: 73–75.

Hossain, M. S., Dey, T. K., Zahid, M. I. and Khan, A. L. (1992). Studies on the Phomopsis leaf blight of arecanut: A new disease in Bangladesh. *Bangladesh J. Plant Pathol.* 8: 1-4.

Iswara Bhat, I. S. (1961). Splitting of arecanut. *Indian Fanning.* 10: 27.

Iswara Bhat, P. S. (1965). Angular planting of areca palms. *Arecanut. J.* 16: 3–5.

Koti Reddy, M. and Saraswathy, N. (1976). Evaluation of certain fungicides in soil against *Ganoderna lucidum* (Leys) Karst pathogenic on arecanut. *Pesticides.* 10: 44–45.

Koti Reddy, M., Saraswathy, N. and Chandramohanan, R. (1978). Diseases of areca nut in India: A review and further considerations. *J. Plantn. Crops.* 6: 28–34.

Lalithakumari, H. (1969). Studies on *Areca catechu* Linn. with special reference to *anoderma lucidum*, a fungal pathogen of areca palm. *PhD thesis*. Microbiology and Pharmacology Laboratory, Indian Institute of Science, Bangalore. p. 161.

Lingaraj, D. S. (1969). Bud rot disease of coconut and arecanut and their control. *Lal Baugh.* 14: 19–21.

Mathai, C. K. (1976). Yellow leaf disease of arecanut: Physiological studies. *Arecanut & Spices Bull.* 8: 33–36.

Nair, R. B. and Rao, K. S. N. (1965). A note on the efficacy of some new fungicides on *Ganoderma lucidum* incitant of anabe in arecanut. *Arecanut J.* 16: 6–8.

Narasimhan, M. J. (1924). Method of preparing casein Bordeaux mixture against supari koleroga. *Mysore Agric. Calender*. pp. 5–8.

Narasimhan, M. J. (1928). A new spraying mixture against areca "koleroga". *Mysore Agric. Calendar for 1927*. pp 24–25.

Nayar, R. (1976). Yellow leaf disease of arecanut: Virus pathological studies. *Arecanut & Spices Bull.* 8: 25–26.

Patel, G. I. and Rao, K. S. N. (1958). Important diseases and pests of arecanut and their control. *Arecanut J.* 9: 89–96.

Ponnamma, K. N. and Solomon, J. J. (2000). Etiology - Phytoplasma. In: *Arecanut Yellow Leaf Diseases*. (Eds. Nampoothiri, K. U. K., Ponnamma, K. N. and Chowdappa, P.). *Technical Bulletin*, No. 39. Kasaragod: CPCRI, pp. 18–23.

Radhakrishnan Nair, R. (1994). Yellow leaf disease of arecanut. In: *Advances in Horticulture*. Vol. 10. Plantation & Spice Crops: Part 2. (Eds. Chadha, K. L. and Rethinam, P). New Delhi: Maltothra Publishing House, pp. 969- 984.

Raghavan, V. and Baruah, H. K. (1956). On factors influencing fruit set and sterility in arecanut (*Areca catechu* L.): l. Studies on pollen grains. *Arecanut J.* 7: 48–56.

Ramakrishnan, T. C. (1990). Control of stem bleeding disease of coconut. *Indian Coconut J.* 20: 13–14.

Ramakrishnan, T. S. (1956). Sun scorch in areca. Proc. *Indian Acad. Sci. Sec. Bull.* 43: 258–263.

Rao, K. S. N. (1960). Plant protection practices in arecanut. *Arecanut J.* 11: 14–16.

Rao, Y. F. and Mohan, S. K. (1976). Bacterial leaf stripe of arecanut caused by *Xanthomonas arecae* sp. *Indian Phytopathol.* 29: 251–255.

Roy, A. K. (1965). Leaf blight of arecanut in Assam. *Arecanut J.* 16: 14–15.

Saidalikutty, K. (1951). Cultivation of arecanut in India. *ICAC Monthly Bull.* 2: 55–57.

Sampath Kumar, S. N. (1985). Epidemiology of bacterial leaf stripe disease of arecanut palm (*HCA catechu* L.) caused by *Xanthomonas campestris* pv. arecae. XI. Influence of management practices on the disease incidence. In: *Arecanut Research and Development.* (Eds. Shama Bhat, K. and Radhakrishnan Nair, C. P). Kasaragod: CPCRI. pp. 124–127.

Sampath Kumar, S. N. (1993). Perpetuation and host range of *Xanthomonas campestris* pv. *arecae* incident bacterial leaf stripe disease of arecanut palm. Adv. *Hort & Forestry.* 3: 99–103.

Sampath Kumar, S. N. and Nambiar, K. K. N. (1990). *Ganoderma* disease of areca palm: Isolation, pathogenicity and control. *J. Plantn. Crops.* 18: 14–18.

Sampath Kumar, S. N. and Nambiar K. K. N. (1993). Relationship among Ganoderma species affecting coconut and arecanut palms, *J. Plantn. Crops.* 21 (Suppl): 119–122.

Sampath Kumar, S. N. and Nambiar, K. K. N. (1996a). Anable disease of arecanut palm: Isolation and pathogenicity tests with *Ganoderma lucidum. Scientific Horticulture.* 5: 63–66.

Sampath Kumar, S. N. and Nambiar, K. K. N. (1996b). Management of anabe (caused by Ganoderma spp) disease of arecanut (*Areca catechu*) chemical and cultural method of control (in India). *Scientific Horticulture.* 5: 57–61.

Saraswathy, N., Koti Reddy, M. and Radhakrishnan Nair, R. (1977). *Colletotrichum gloeosporioides* causing inflorescence dieback, button shedding and nut rot of betel nut palm. *Pl. Dis. Rep.* 61: 172–174.

Sarma, B. K. and Murthy, K. N. (1971). Crown rot of arecanut (*Areca catechu* Linn.] *Assam. Sci. Soc.* 14: 107–110.

Vinayaka Hegde and Gunasekaran, M. (2004). Management of basal stem rot disease *of HCA catechu* L. in India. *J. Plantn. Crops.* 32: 25–27.

CHAPTER 2

PRESENT SCENARIO OF BETELVINE OR *PAAN* (*PIPER BETLE L.*) DISEASES AND THEIR MANAGEMENT

J. N. SRIVASTAVA[1*], A. K. SINGH[2], A. K. SINGH[3], and S. R. P. SINGH[4]

[1]*Department of Plant Pathology, Bihar Agricultural University, Sabour 813210, Bhagalpur, Bihar, India*

[2]*Division of Plant Pathology, Sher-e-Kashmir University of Agricultural Sciences and Technology of Kashmir (SKUAST-J), Chatha, Jammu 180009, J & K, India*

[3]*Department of Agronomy, Rajendra Prasad Central Agricultural University, Pusa, Samastipur, Bihar, India*

[4]*Department of Horticulture, College of Horticulture, Noorsarai, Nalanda 803113, Bihar, India*

[*]*Corresponding author. E-mail: j.n.srivastava1971@gmail.com*

ABSTRACT

Betel vine (*Piper betle* L.) is basically a plantation crop that belongs to the family Piperaceae. Betelvine plants are evergreen, dioecious, perennial climber, and cultivated for its leaf. Betel vine is a shed-loving plant and originated from Malaysia (Chattopadhyay and Maiti. 1967. *Diseases of betelvine and species*. New Delhi: ICAR). Betel vine is known by different names in India and abroad. In India, Betel vine is known as *Paan* in Hindi and Bangala, Tambula in Sanskrit, Villayadela in Kannada, Vettilakkoti in Malyalam, Vettilai in Tamil, Tamalapaku in Telugu, Videch-Paan in Marathi, and Nagarbel in Gujarati. Historically, the word *Paan* in Hindi and other Indian languages is probably a derivative of the Sanskrit word *Paan* meaning leaf. In foreign languages, it is known as Burg-e-Tanbol in Persian and Tanbol in Arabic. Betel vine is cultivated in many countries, namely Bangladesh, Malaysia, Sri Lanka, Pakistan, Singapore, Thailand, Myanmar, and India. In India, this plant is commercially cultivated in Andhra Pradesh, Assam, Bengal, Bihar, Karnataka, Kerala, Madhya Pradesh (M.P.), Maharashtra, Orissa, Tamil Nadu, Tripura, and Uttar Pradesh. Betel vine is a rich source of vitamins A and C, 21 essential oils, alkaloids, arecoline, etc. Besides this, it is supposed to be a tonic for the brain, lever, and heart. The leaves of this plant are used for chewing traditionally from ancient times. It is also offered after lunch and dinner and also during other social gatherings. In fact, no Hindu religious ceremony is complete without *Paan*. The yield of leaves of betel vine is highly reduced due to the occurrence of several fungal, bacterial, and viral diseases. Leaf rot disease and leaf spot disease caused by pathogens, *Phytophthora parasitica* and *Colletotrichum capsica*, are the major constraints for the cultivation of the crops across the country (Goswami et al., 2002). Some of the important diseases encountered in betelvine plantation are discussed in this chapter.

2.1 FUNGAL DISEASES

2.1.1 PHYTOPHTHORA FOOT ROT/PHYTOPHTHORA LEAF ROT

Introduction and Economic Importance

Phytophthora foot rot and *Phytophthora* leaf rot diseases occur in all betelvine-growing countries, namely Bangladesh, Malaysia, Sri Lanka, Pakistan, Singapore, Thailand, Indonesia, Myanmar, and India. In India, foot rot and leaf rot diseases of betel vine were first reported by Dastur in 1927 from Durg, M.P., India. The highest incidence of foot and leaf rot diseases has been also reported from Midnapore and Nadia districts of West Bengal (Dasgupta and Sen, 1999) and other states of India. The yield of losses (leaves) varies from 30% to 100% in the case of foot rot disease and leaf rot disease, leading to almost total crop failure (Maiti and Sen, 1979a; 1982; Dasgupta et al., 2000).

Symptoms:

The most common symptoms associated with foot rot are wilting condition. The wilted plants contrast sharply in comparison to healthy plants by their stark, chlorotic appearance, and shriveled leaves; initially, the leaves slight drop further, leading to dropping effects on the whole canopy. Fine young roots of diseased plants are infected first and caused rot symptoms. Gradually, the rotting spreads through older roots, and ultimately rotting symptoms reach the foot or collar region of the plant. Finally, the whole underground portion gets more or less completely rotten. The soft tissues of old roots and the inter-nodal portion of the cuttings are completely decomposed by the pathogen, leaving only the fibrous portion of roots (Maiti and Sen, 1979a).

Leaf rot by pathogen can damage the crops within a week when the disease spreads to the vine (Chaurasia, 1994). The leaf rot disease symptoms lead to circular black-colored or brownish-colored water-soaked spots. The size of spots rapidly increase and coalesce with each other, involving a major area in the leaf blade, which undergoes rotting when the weather is continuously wet. Maiti and Sen (1977) observed two types of symptoms in betelvine plants in West Bengal. In Type-I symptoms, the expending spots are circular, necrotic deep brown in color with a distant gray-brown zonation developed, when the relative humidity fluctuates greatly during day and night. While in Type-II symptoms, the expending spots are circular, necrotic dark brown in color, and no zonation develops when the weather is continuously wet. Infection that mainly occurs on the leaves is located within a couple of feet from the ground surface, while the upper leaves are free from rotting.

Leaf rot and foot rot caused by *Phytophthora palmivora* losses of leaf yields range 30%–100% (Maiti and Sen, 1997).

Causal Organism: *Phytophthora parasitica* var. *piperina* Dast. *Phytophthora palmivora.*

Pathogen:

The fungus produces hyaline, coenocytic mycelium. The sporangia are thin walled, hyaline ovate, or learn shaped with papillae, measuring 30–40 × 15–20 μm. Zoospores liberated from the sporangia are kidney shaped, and in biflagellate zoospores, one flagellum is whiplash and another is tinsel type. Oospores are dark brown in color, globose, with thick-walled produced after sexual reproduction.

Disease Cycle:

The pathogen is soil-borne and survives as facultative saprophyte nature in the infected plant debris and in the oospores form in the soil. The fungus is disseminated from one field to another field through irrigation water. The secondary inoculums, sporangia, and zoospores spread through splash, irrigation, and wind-borne rains.

Epidemiology:

Just after the rainy season, disease becomes more severe. September to February are most favorable months for disease development because during this period, atmospheric humidity is high and night temperature is low (<23 °C) which are conducive for disease progress.

Management:

The leaf rot disease is more contagious; hence, it is very difficult to control once a crop is infected. Therefore, during the rainy season, one needs to take more care of the crop. Below-mentioned steps should be followed to manage the leaf rot disease:

1. Use disease-free planting materials of betel vine.
2. Dip the betelvine's seedling/cutting in a Streptocycline 500 ppm + Bordeaux mixture 0.05% solution for 30 min before transplanting.
3. At the time of planting, the seed vines/setts should be dipped in the Bordeaux mixture 1 per cent solution or the Ceresan 0.1% solution at least for 20 min.
4. During the selection of site for the cultivation of betel vine, select a well-drained area for Boroj (a special structure for the betel vine cultivated area).
5. When symptoms appear, dig up the infected betel creeper and burn it.
6. Drench vines with Bordeaux mixture (0.5%) or 0.2% Zineb (Dithane Z-78) on the ground near vines and also spray the Bordeaux mixture on diseased leaves at an interval of 15 days from the last week of June to October.
7. Drench betel vines with the Bordeaux mixture (1 copper: 1 lime: 1 water) during the wet weather for new plants to reduce the incidence of disease.
8. Spray metalaxyl with 2.5 g/L water on the leaves of betel vines at the intervals of 15 days.

2.1.2 LEAF SPOT/MARGINAL BLIGHT/ANTHRACNOSE DISEASE

Introduction and Economic Importance

The leaf spot disease of betel vine was first identified by Roy (1948) in Bangladesh. In India, anthracnose was first described by Hector (1927) from West Bengal. But the anthracnose disease of betel leaf caused by

Collelotrichum capsica (*C. capsici*) (Syd.), E. J. Butler & Bisby, or *Collelotrichum piperis*, Petch, is the prime constraint for the cultivation of betel across India. Four different fungi have reported from India causing anthracnose. The disease is fairly widespread and makes its appearance compulsory if there are occasional showers in April. Maiti and Sen (1982) reported that *C. capsici* is of major importance in West Bengal and may cause loss up to 25%–90%. Singh and Joshi (1972) also recorded extensive damage of *C. capsici* in M.P. during July–September.

Symptoms:

Anthracnose symptoms on the leaves are characterized by the presence of spots which are usually circular but sometimes irregular in shape and size with brownish black centered with a yellow halo. These leaf spots rapidly increase in size and girdle the stem culminating in the death of the betel vine. These spots often coalesce to form large lesions. The infected regions gradually become thin and dry without any rotting. When the spots are present on the margin of the leaves, the leaf blade tends to droop due to the shrinkage of tissues. The infected leaves may fall off prematurely. Acervuli are produced in abundance on aerial stems and branches, especially on the leaves. Marginal leaf tissue becomes black in color, necrotic, and gradually spreads toward the leaf center (Basak et al., 1992).

On the stem, disease is characterized by as small, black, circular specks that appear under the green bark of the stem. These specks, if conditions are dry, usually do not increase in size and remain as a black stain on the surface of the stem. Occasionally, the bark in such cases may dry up and rupture, exposing the wood underneath (Bhale et al., 1987).

Causal Organism:

1. *C. capsici* (Syd.) Butler & Bisby
2. *C. piperis* Petch
3. *C. dasturi* Roy
4. *Glomerella cingulata* (Stonem) Spauld & Schrenk

Maiti and Sen (1979b) reported that the actual pathogen of leaf spot/ marginal blight/anthracnose disease is *C. capsici.*

Colletotrichum Capsici:

The mycelium is hyaline, extrametrical, turning dark gray to black, and develop acervuli. Acervuli are dark, brownish black, erumpent, 35–105 μm in diameter. Conidiophores are simple, hyaline, nonseptate, 22.5–67.0 × 2.3–3.4 μm and are produced on the surface of the stromatic tissues. The conidia are terminal, unicellular, hyaline, subcylindrical, and measure 11.6–19.8 × 3.1–5.9 μm. Setae are dark brown, septate, tapering, and measure 85.3–186.2 × 3.2–5.6 μm.

Colletotrichum Piperis:

In vitro hyphae are first bud off conidia from their tips, but later they organize into a definite stroma, brown to olivaceous in color, with a tuft of elongated cells, which cut off conidia from the tips. The setae are many-celled, apical cells. These are hyaline or lighter than the rest and measure 40–120 × 6.0 μm. The conidia are elliptical with rounded ends and a central vacuole; hyaline may be brown in old cultures, measuring generally 12.5–15.0 × 5.0 μm (range 10.0–16.2 × 3.7–6.2 μm).

Colletotrichum Dasturi:

Acervuli are distinctly erumpent and appear as tufts with a narrow base on the epidermis. The stroma, which consists of pseudo-parenchymatous cells, is limited to a few cells in the epidermis by a narrow opening. Acervuli are setose and brown. Each seta is swollen at the base and gradually tapers from the base toward the apex. The setae are 110.4–360.0 μm in length and 3.4–4.6 μm in breadth, with the average dimension being 136.7 × 4.2 μm. The conidiophores are simple, straight, and closely packed in the center of acervuli. Conidia pointed at both ends, occasionally linear and· straight, hyaline, or light brown, measuring 19.4–25.1 × 2.3–3.4 μm (average dimension 32.1 × 3.0 μm).

Glomerella Cingulata:

The morphological characters of the fungus on artificial media have been studied by McRae (1934) and Chowdhury (1945). The acervuli are round *or* oblong black or *pink* and with or without setae. The conidiophores measure 25–30 μm. The conidia are hyaline or pink in mass, straight, rounded at the ends and measure 11–32 × 4.0–55 μm and 8.419.0 × 3.5–6.4 μm, with the average dimension being 13.4 × 4.6 μm. Perithecia when formed are globose, ostiolate, dark brown, measuring 99.8–236.1 × 118.8–274.3 μm in dimensions. The ostiole is short or long, papillate, light colored, and measures 30–150 μm. The ascus is 8-spored; paraphyses are absent. The ascospores are hyaline, unicellular, oblong to slightly fluxoid, round at both ends, measuring 14.0–18.0 × 6.8–11.0 μm (McRae, 1934).

Disease Cycle:

Pathogens persist in the infected plant debris in the field. The primary infection is through soil-borne conidia. These conidia are disseminated by rain splash or through irrigation water. The secondary infection in the field is through air-borne conidia.

Epidemiology:

The infection rate gradually attains maximum when average temperature reaches 28 °C, relative humidly becomes more than 90% with rainfall at frequent intervals. Roy (1948) recorded severe leaf spot of betel vine due to high rainfall. Maximum disease severity is observed in the last weeks of June and July months when all the three factors are higher than in other months of the year. However, the incidence of leaf spot is also observed during March to May (Goswami et al., 2002).

Management:

1. Collect all infected plant debris and destroy them.
2. Spray Bavistin @ 0.1% or Captan @ 1.5% or mancozeb @ 0.25%, or copper oxychloride @ 0.3% as soon as the disease appears and at the 12-day interval.
3. Spray Ziram @ 0.2% or Bordeaux mixture @ 0.5% after plucking the leaves.

2.1.3 POWDERY MILDEW DISEASE

Introduction and Economic Importance:

The disease was first reported in the world from Ceylon by Stevenson (1926) and later from Myanmar (formerly Burma) by Mitra (1930). In India, the disease was first observed in Mysore by Narasimhan (1933). The disease was also reported from Mumbai (formerly Bombay) by Uppal and Kamat (1938). It was later observed from almost all betelvine-growing areas of India (Jhamaria and Daftari, 1970; Nema and Nayak, 1975; Maiti, 1977). Rao (1993) reported high yield losses ranging 70%–90% due to this disease.

Symptoms:

The disease attacks the crop at any stage of its growth and all above the ground portion of the plant. But infection is mainly noticed on tender shoots and leaves of the plant. Whitish dusting growth like powder appears on both the surfaces of leaves which later spreads and covers the main portion of the leaves. The affected tender shoots and buds are deformed and shriveled, and the margins of leaves turn inward. Young leaves after infection become deformed, with their surfaces being cracked and the growth checked. Such leaves become pale in color and sometimes may drop with slight disturbance. The disease is more serious in old plantations (Maiti, 1977). In the advance stage of the disease, whitish dust turns to dark brown due to the formation of cleistothecia when there are unfavorable environmental conditions for the growth of the plant, which are otherwise favorable for disease development.

Causal Organism: *Odium piperis.*

Pathogen:

The fungus is an ectoparasite and obligate in nature. It produces hyaline, septate and profusely branched hyphae on the surface of the leaves, and intakes nutrient through haustoria. Conidiophores are aseptate, hyaline, short, and club shaped. They produce conidia in chains and are arranged in a basipetal manner. Conidia are single celled, hyaline elliptical, and borne over short conidiophores.

Disease Cycle:

The pathogen persists in the infected plant debris in the soil. The primary infection is from the soil-borne resting body. The secondary infection in the crops is through wind-borne conidia which spreads through rain splash or irrigation water.

Epidemiology:

The disease incidence gradually attains maximum when temperature reaches 26 °C, relative humidly becomes more than 90%, and there is a dry humid weather during the months of May–July.

Management:

1. Maiti (1994) reported betelvine cultivars such as male clones of Kapoori, Tellaku, Vellairettala, Ambadi Badam, and Kulgedu were found resistant to powdery mildew.
2. Can be controlled with two preventive sprays with wettable sulfur @ 0.3% or Dinocap (Karathane) @ 0.05% from the initial appearance of the disease. Sprays should be repeated at 10–15 days.
3. Dusting twice or thrice with fine sulfur will check the disease.

2.1.4 SCLEROTIAL WILT/SCLEROTIAL FOOT ROT/STEM ROT/ COLLAR ROT/BASAL ROT

Introduction and Economic Importance:

Sclerotial wilt/sclerotial foot rot/stem rot/collar rot/basal rot of betel vine (*Piper betle* L.) caused by *Sclerotium rolfsii* (*S. rolfsii*) is prevalent in almost all the betelvine-growing areas of India (McRae, 1930). The disease was first investigated by Shaw and Ajrekar (1915). They reported a causal organism under the name of *Rhizoctonia destruens*. Later, Hector (1927) discovered a mix population of *S. rolfsii* and *Phytophthora* associated with the foot rot. McRae (1928) established the parasitism of *S. rolfsii* isolated from the betel vine in West Bengal.

The betel vine crop is cultivated in artificially prepared gardens. Locally, it is called "Bareja." The structure of bareja provides inside moist, humid and shaded condition is favorable for vine growth these microclimatic conditions are also conducive for development of many root and aerial diseases such as sclerotial wilt, sclerotial foot rot, stem rot, collar rot, basal rot, and leaf rot. They estimated the losses of 25%–90% due to this disease.

Symptoms:

Betelvine crops are susceptible at all stages of the magnification to the disease. The characteristic symptoms of the disease are darkening of the stem at the foot (near the ground level) of the plant. Infection caused by *S. rolfsii* conventionally begins at the collar region of the susceptible host, as white, cottony lesion, which later forms sclerotia in the collar region of dead plants (Punja and Grogan, 1985). The darkened portions of the stem become shrinked, soft, and turn black. Pathogens are developed white ropy, fan-shaped mycelial strands on the infected portions of the stem. Brown to dark brown sclerotia appears on the infected portion of the plants. The leaves turn yellow, become flaccid, and droop off. Ultimately, the whole betelvine plants wilt and dry up. Even after the death of the plants, sclerotia are present in the collar region of the dead plants.

Causal Organism: *Sclerotium rolfsii*

Pathogen:

The *S. rolfsii* produces profuse branched mycelium of white to gray color. Sclerotia are spherical, smooth, and shiny brown-colored-like mustard. They are seen on the infected stem and collar region.

Disease Cycle:

The pathogen is soil-borne and grows saprophytically on the crop debris or in soil. It also survives as sclerotia in the infected plant debris in the soil for more than one year. The sclerotia are disseminated through irrigation water.

Epidemiology:

May to July are the most favorable months for disease development when temperature ranges from 28 °C to 30 °C with high atmospheric humidity.

Management:

1. Remove the affected betelvine plants along with the roots.
2. Apply soil amendments using neem cake, mustard cake, or farmyard manure in soil.
3. Use the recommended dose of nitrogenous fertilizer.
4. Drench with carbendazim @ 0.1% in the soil.
5. Apply pentachloronitrobenzene (Brassicol) fungicide should be done @ 15–20 kg/hectare for effective management.
6. Spray fungicide like copper oxychloride @ 0.3 g/L water or carbendazim @ 0.1 g/L water, 2–3 times at 15-day intervals but first spray that on the appearance of the disease.

2.1.5 RHIZOCTONIA FOOT ROT

Introduction and Economic Importance:

Rhizoctonia bataticola (*Macrophomina phaseolina*) is a plant pathogenic fungus with a wide host range and worldwide distribution. *Rhizoctonia bataticola* is the best known pathogen that causes various plant diseases such as collar rot, foot rot, and root rot (Latorre, 2004). *Rhizoctonia bataticola* is one of the most devastating seed- and soil-borne pathogens. This disease is severe in many parts of India, especially in Punjab and Gujarat states.

Symptoms:

Initial symptoms seem similar to foot rot infection caused by *Phytophthora* or *Pythium*. Characteristic symptoms of the disease are the appearance of reddish-brown-decaying lesions on the stem base and on the outer cortical layer of the older roots. In the advance stage of disease, the girdling lesion on the stem base, at or below the soil level, becomes discolored, and the rotten portion is distinctly dry unlike that of the foot rot. Such lesions usually girdle the stem and appear as a canker thereon. The disease is more severe during postmonsoon months (Dasgupta and Maiti, 2008).

Causal Organism: *Rhizoctonia bataticola.*

Pathogen:

The mycelium of the fungus is septate and fairly thick, and produces black-colored, irregular sclerotia which measures 100 μm in diameter.

Disease Cycle:

The fungus *Rhizoctonia bataticola* survives in the soil as a saprophyte in the form of sclerotia for many years. The sclerotia are disseminated through irrigation water, heavy winds, and other cultural operations.

Epidemiology:

Dry weather conditions after heavy rains, high soil temperature (35 °C–39 °C) and low soil moisture (15%–20%) favored the disease development.

Management:

1. Deep summer ploughing.
2. Clean cultivation and removal of infected plant parts usually reduce the disease inoculum.
3. Destruction of dead vines.
4. Use of organic manures reduces the severity of the disease.
5. Application of any copper fungicide like copper oxychloride of 0.3% concentration at the rate of 5 L/m for effective management.
6. Diseased patches of crops drench with carbendazim @ 0.1% or benomyl @ 0.05%.

2.1.6 *FUSARIUM WILT*

Introduction and Economic Importance:

Fusarium wilt is the most destructive soil-borne disease throughout the world. This pathogen blocks the xylem vessels. As a result, the infected plants become unable to transport water and nutrients from soil, resulting in severe wilt and death of plants. The pathogen is difficult to control because of its soil-borne nature and wide host range. A combination of high temperature and poor drainage favors disease development.

Symptoms:

The characteristic symptoms of fusarium wilt are yellowing of older leaves like chlorosis, necrosis, and premature leaf drop, wilting, results browning of the vascular system, stunting of growth, and ultimately drying and death of the plant.

Causal Organism:

Several species of *Fusarium* have been isolated from the wilted plants of betel vine. These species are generally considered earlier as saprophytic like *Fusarium equiseti* (*F. equiseti*) (Corda) Sacco (Singh and Joshi, 1972) and *Fusarium semitectum* (*F. semitectum*) Berk. & Rav. (Mathur and Sinha, 1959) except two parasitic forms *Fusarium moniliforme* (*F. moniliforme*) (Sheld.) Snyder & Hansen and *Fusarium oxysporum* (*F. oxysporum*) (Singh and Joshi, 1972). Pathogenicity tests in all the earlier reports have not been convincing.

F. equiseti *(Corda) Sacco:*

Conidia are falcate with a well-developed, pedicellate base cell, and an attenuated apical, cell which are bent inward, exaggerating the normal curve of the spores. Mature conidia have 4–7 thin but distinct septa and measure 22–60 × 3.5–5.9 μm. Chlamydospores are intercalary, solitary, in chains or in knots, globose, 7–9 μm in diameter.

F. oxysporum *Schlecht:*

Microconidia are borne on simple phialides arising laterally on the hyphae and from short sparsely branched conidiophores. Microconidia are generally abundant, variable, oval-ellipsoid, cylindrical, and straight to curved 5–12 × 2.2–3.5 μm. Macroconidia are borne on more elaborately branched conidiophores or on the surface of tubercularia like sporodochia. They are thick walled, generally 35 septate, fusoid to subulate, and pointed at both ends. Occasionally, fusoid to falcate, macroconidia are found with a somewhat hooked apex and a pedicellate base, measuring 27–66 × 3–5 μm. Chlamydospores, both smooth and rough walled, are generally abundant and are both terminal and intercalary, usually solitary but may form a pair or chain.

F. moniliforme *(Sheldon) Snyder & Hansen:*

Macroconidiophores are simple, lateral, subulate phialides formed on the aerial hyphae; they rarely form on short lateral branches. They are 20–30 × 2.3 µm. Microconidia are formed in chains under optimum growing conditions. They measure 5–12 × 1.5–2.5 µm and are fusiform to clavate with a slightly flattened base, occasionally single septate. The formation of macroconidia is rare. Conidiophores consist of a single basal cell bearing 2–3 apical phialides or it may form 2–3 metulae which bear filiform or obclavate phialides. They measure 20–24 × 3.5–4.0 µm. Macroconidia are inequilaterally fusoid, delicate, thick walled, with an elongated, often sharply curved apical cell and pedicellate basal cell. They are 37 septate and measure 25–60 × 2.5–4.0 µm. Chlamydospores are absent.

F. solani *(Mart.) Snyder & Hansen:*

Microconidiophores are elongated and sparsely branched, reaching up to 400 µm in length. Each branch of microconidia usually terminates in a single cylindrical to barely subulate phialide which measures 45–80 × 2.5–3.0 µm. Microconidia are broader and more oval in shape with somewhat thicker walls that measure 8–16 × 2.4 µm and may be single septate. Macroconidia are inequilaterally fusoid with many of the spores having a rounded foot cell, macroconidia measure 35–100 × 4.6–8.0 µm. Chlamydospores develop abundantly on weak media. They are globose to oval, smooth to rough walled, and measure 9–12 × 8–10 µm.

Disease Cycle:

The pathogen persists in the soil or crop debris as a mycelium and all spore types. But it is most commonly recovered from the soil as chlamydospores. This pathogen disseminates in two ways: it disseminates short distances by water splash or irrigation water, and by planting equipment, and long distances by infected transplants and vine sets. Secondary infections are by conidia through rain or wind.

Epidemiology:

High temperature and high relative humidity favor the disease development.

Management:

1. Cultural operation like deep summer ploughing, crop rotation with nonhosts will help to manage the disease.

2. Select wilt resistant varieties for transplanting.
3. Sett treatment with 4 g *Trichoderma viride* or 2 g carbendazim seed before transplanting is effective.
4. Drenching soil with the Bordeaux mixture @ 1% or blue copper @ 0.25% may protect the plant.

2.1.7 FUSARIUM LEAF SPOT

Introduction and Economic Importance:

Fusarium leaf spot disease was reported from several betelvine-growing areas of Uttar Pradesh by Mathur and Sinha (1959). *Fusarium* leaf spot disease is caused by *F. equiseti*. Further, this leaf spot disease, caused by *F. semjtectum*, was reported from West Bengal by Chattopadhyay and Sengupta (1955). Hector (1924) stated that species of *Fusarium* could be isolated along with other fungi from betel vines, showing yellow-colored discoloration and wilting of leaves and blackish-brown decay of roots.

Symptoms:

Plants can be infected at any stage of their development. The disease initially appears on the leaves as large circular spots with a number of board and distinct concentric zones of alternate chocolate brown and light brown color. Plants show the yellowing of leaves first. Plants wilt gradually. Often sudden withering and drying of the entire plant take place. The infection proceeds upward through the vascular system, causing the blockade of water transportation and finally wilting starts. The decayed area is water soaked, slightly sunken, but is still firm often with white moldy growth on the lesion.

Symptoms produced by *F. semjtectum* on infected flesh are almost the same as produced by *F. equiseti*. But *F. semjtectum* shows faster symptoms and decays the infected tissue within six days.

Causal Organisms:

1. *F. equiseti* (Corda) Sacco.
2. *F. semitectum* Berk and Rav. Syn. *F. pallidoroseum* (Cooke) Sacco.
3. *F. equiseti* and *F. semitectum* are polyphagous fungi. These are pathogenic on various crops. *F. semitectum* and also have a broad host range.

F. Equiseti:

Mycelium, the vegetative part of fungus, is hyaline, branched, septate, measuring 4.05.6 μm. Conidia are falcate with a well-developed, pedicellate base cell, and an attenuated apical, cell which are bent inward, exaggerating the normal curve of the spores. Mature conidia have 4–7 thin but distinct septa and measure 22–60 × 3.5–5.9 μm. Chlamydospores are intercalary, solitary, in chains or in knots, globose, measuring 7–9 μm in diameter.

F. Semitectum:

Mycelium, the vegetative part of fungus, is hyaline, septate, and branched. Sporodochia-lacking conidia are formed on the aerial mycelium from loosely branched conidiophores. Each branch is terminated in a conidiogenous cell which appears to form a conidium from a single apical pore and then forms successively a second, third, or even fourth pore, thus forming a polyblastic, sympodial cell. The conidia were straight or slightly spindle shaped. Smaller conidia aseptate to 2-septate, larger usually 3-septate, measuring 26.42–43.68 × 2.9-6.0 μm seldom 4-, 5-, or 7-septate.

Disease Cycle:

The pathogen persists in the soil debris as a mycelium and in all spore types. But it is most commonly recovered from the soil as chlamydospores. This pathogen disseminates in two basic ways. It disseminates short distances by water splash or irrigation water, and by planting equipment, and long distances by infected transplants and vine sets. Sources of secondary infection are conidia which spread through rain or wind.

Epidemiology:

Moderate temperature and high relative humidity favor the development of disease. Temperature of 20 °C or above is favorable for the growth of *F. equiseti* and *F. semitectum*. Temperature of more than 30 °C supports the maximum growth of *F. equiseti.*

Management:

1. *Fusarium* leaf spot Can be controlled with two preventive sprays with copper oxychloride @ 0.3% from the initial appearance of the disease and repeat at 10–15 days of intervals.

2.1.8 CERCOSPORA LEAF SPOT

Introduction and Economic Importance:

The disease was first reported from Uttar Pradesh in 1971 and subsequently from West Bengal in 1977 (Maiti and Sen, 1979b). The loss from this disease is estimated to be 10%–20% in severely infected plantations (Maiti, 1977).

Symptoms:

First symptoms are noticed on the leaves as minute, deep-brown to black dots, developing close to each other in clusters. The surrounding portion develops a yellow halo. The yellow halo portion is usually more than the black-dotted portion. The symptoms can appear in any part of the leaves, but more frequently, they appear toward the margin and tips. The older leaves show more susceptibility than the younger ones. Premature shedding of infected leaves occur owing to severe infection.

Causal Organisms: *Cercospora piperis betle* Sawada & KaLSuki

Pathogen:

Conidiophores are nonfesciculate, septate, nongeniculate, olive-brown to light brown in color, measuring 22.6–78.0 × 2.8–5.6 μm. Conidia subhyaline, straight to curved, 5- to 15-septate, truncate at the base *with* acute tips, 26.8–152.8 × 2.2–3.9 μm.

Disease Cycle:

The primary source of inoculum is infected plant debris. The source of secondary infection is air-borne conidia produced on leaves.

Epidemiology:

Low temperature and high humidity favor the infection.

Management:

1. Destruct infected plant residues.
2. Spray mancozeb @ 0.2% or carbendazim @ 0.1% twice fortnightly interval, starting with disease appearance.
3. Spray copper oxychloride @ 0.3% or zineb @ 0.2% or Vitavax @ 0.1% at the initiation of the disease.

2.2 BACTERIAL DISEASES

2.2.1 *PSEUDOMONAS BACTERIAL LEAF SPOT/BACTERIAL BLIGHT*

Introduction and Economic Importance:

The disease was first reported from Sri Lanka by Raghunathan (1926). Later, Raghunathan (1928) gave a detailed account of the disease, and identified and described the organism as a new species of *Bacterium betle* Raghunathan. This bacterium has now been placed under *Pseudomonas* and renamed as *Pseudomonas betlis* (Raghunathan) Burkholder (Breed et al., 1948). Asthana and Mahmud (1944) observed the occurrence of this disease in M.P. where it was observed to cause a good amount of loss. This disease popularly known as "Kinrog" in M.P. is believed to have been present for long time, but it could fetch partial attention. It has also been reported from Mauritius.

Symptoms:

Symptoms occur on the lower surface of the leaves as minute, water-soaked spots develop, and within 2–3 days, the disease starts appearing on the upper surface as brown to dark-brown circular to angular spots surrounded by a yellow halo zone. The center of the spots presents a cracked mottled appearance. Under dry conditions, the infection remains localized and measure 1–5 mm in diameter. However, under humid conditions, they enlarge and may measure up to 1 cm and coalesce forming large irregular necrotic areas. Leaves thus affected lose their luster, become flaccid, turn yellow, and finally drop off. Under humid conditions, disease may become severe and ooze containing bacteria secretes from the lower surface of leaves. Raghunathan (1926) reported the disease on stems, but Asthana and Mahmud (1944) observed it exclusively on leaves close to the soil surface, up to a height of 33 cm in dry weather and up to 1 m in wet weather.

Causal Organisms: *Pseudomonas betlis* (Raghunathan) Burkholder.

Pathogen:

The organism is cylindrical, short-rod-shaped, with slightly rounded ends, measuring 1.52.5 × 0.5–0.7 μm in single, pairs, or short chains. It is nonmotile, noncapsulated, nonsporulating, gram negative, and aerobic. Good growth occurs in bovril broth; the organism produces pellicles. Little growth occurs in Cohn's or Uschinsky's solutions, and none in Fermi's acid. No gas formation takes place in the lactose, maltose, or sucrose medium. Starch is hydrolyzed and gelatin is liquefied and turned green. Pathogen can resist desiccation for three days only (Raghunathan, 1928).

Disease Cycle:

Bacteria which are viable in the infected betelvine leaves serve as a primary source of inoculum. Splashes of rain and irrigation water help in the secondary spread. Asthana and Mahmud (1944) reported that the disease is more prevalent during the monsoon season in heavy soils and in low-lying areas, water-logged plantations. The disease is disseminated by rain, water, insects, and mites. The intensity of the disease has a direct relationship with insect incidence.

Epidemiology:

Cloudy weather with intermittent rains and high relative humidity are favorable conditions for disease development. Two-to-three-year-old vines are highly susceptible.

Management:

1. Pull out and bury the infected vines in the field.
2. Regulate irrigation in crops during the cold weather season to control the disease.
3. Spray with the solution of the Streptocycline 400 ppm + Bordeaux mixture @ 0.25% at 20-day intervals, after plucking the leaves.

2.2.2 XANTHOMONAS BACTERIAL LEAF SPOT/BACTERIAL BLIGHT

Introduction and Economic Importance:

Xanthomonas bacterial leaf spot/bacterial blight was first time reported by Patel et al. (1951) who identified the pathogen as *Xanthomonas betlicola*. This was considered as a minor disease, even though there were subsequent reports of its occurrence in 1971 from the Jabalpur area of M.P. (Singh and Chand, 1971) and in 1978 from Kerala by Mathew et al. (1978). Mahesha et al. (2009) reported up to 60% incidence of this disease from major betel-cultivating areas, and it can increase to about 75% during the rainy season.

Symptoms:

Symptoms appear initially as tiny, brown, water-soaked specks on the leaves surrounded by a yellow halo, which later enlarge and become necrotic and angular, mostly confined to interveinal areas. Infected leaves lose their luster, turn yellow, show withering, and ultimately fall off. Under wet weather conditions, infection goes to stem and shows small, elongated black lesions on lower nodes and internodes. These lesions increase both longitudinally and horizontally in size, and blackening may be extended to the length of several nodes. The stem tissues become weak and break easily at the infected nodes and the betelvine plants showed withering and drying.

Causal Organisms: *Xanthomonas campestris var betlicola.*

Bacterium is a small rod with a single polar flagellum. It is gram negative and nonspore forming.

Disease Cycle:

The bacteria persist in the infected betelvine leaves, and infected leaves serve as a primary source of inoculum. Rain splashes and splash irrigation water help in the secondary infection.

Epidemiology:

Cloudy weather with intermittent rains and high relative humidity are favorable conditions for disease development. Two-to-three-year-old betelvine plants are highly susceptible.

Management:

1. Remove and burn the infected betelvine plants.
2. Regulate irrigation during the cold weather season to control the disease.
3. Spray the solution of the Streptocycline 400 ppm + Bordeaux mixture @ 0.25% at 20-day intervals, after plucking the leaves.

2.3 VIRAL DISEASES

2.3.1 BETELVINE MOSAIC DISEASE

Introduction and Economic Importance:

Betelvine mosaic disease was first time reported by Rao et al. (1986) and Singh (1986) after observing it in the experimental garden of *All India Coordinated Research Projects on betel vine*, Chintalapudi (Ponnur), Andhra Pradesh. The local cv. Ponnur-S was found to be highly susceptible to the disease. Subsequently, it was reported from M.P. by Deshpande (1986) in cvs. Kalkatia and Bilhari.

Symptoms:

The most prominent symptom of the disease expressed in susceptible cultivars was the development of the typical mosaic pattern on the leaf lamina. The disease starts with mild vein-clearing and formation of chlorotic lesions in the interveinal area which in advanced stages led to the development of the characteristic mosaic pattern. The leaves rolled upward from the margin when the plants were infected in the early stages of growth. Puckering and blistering of the leaf blade are also seen in affected leaves in the case of severe infection. Small size of leaves (little leaf condition), upward rolling of leaf margins, and retardation of growth of affected plants were also common. At acute stages of infection, the emerging leaves became completely yellow and reduced in size. Masking of symptoms was noticed in the infected plants during the months of December and January when the temperature was low. Such leaves became thick and shiny. But when temperatures slightly decreased, the earlier symptoms appeared again in the new flush. Symptom expression also varied from cultivar to cultivar, depending upon the degree of susceptibility. Highly susceptible cultivars produced symptoms typical to those described earlier, whereas tolerant and resistant cultivars showed relatively mild mosaic symptom (Deshpande, 1986).

Causal Organisms: Betelvine mosaic disease.

The causal agents are identified as the flexuous rods of 730 × 12 nm in size having a protein subunit with a molecular weight of 32,000 KD (NBRI 1987-88). On the basis of the above studies, they classified the agents under the potyvirus group.

Disease Cycle:

Betel vine is propagated through stem cuttings; therefore, the primary source of inoculums might come through the planting material and might be transmitted further within the garden by either physical contact of the plants or during cultural operations.

Transmission:

The agent was reported to have been transmitted from diseased to healthy plants by grafting approaches (Deshpande, 1986; Singh, 1989) and by mechanical sap transmission (Rao, 1987). Transmissibility of the virus by leaf patch grafting was also reported by Singh (1989).

Epidemiology:

Various weather conditions and soil factors affected the incidence and spread of mosaic disease. Maximum temperature and soil moisture were positively correlated with the incidence of the disease, while minimum temperature and relative humidity did not have any influence. pH had negative correlation in one garden.

The rate of spread of mosaic disease was maximum during initial stages (October and November) and gradually decreased during winter (December and January).

Management:

1. Grow resistance varieties.
2. Control the insect vector.

KEYWORDS

- **Blend Psyllium**
- **etiology**
- **symptomatology**
- **management**

REFERENCES:

Asthana, R. P. and Mahmud, K. A. (1944). Bacteria! leaf spot of *Piper betle*. *Indian J. Agric. Sci.* 14, 238–288.

Basak, A. B., Mridha, M. A. U., and Jlali, M. A. (1992). Studies on the leaf spot disease of *Piper betel,* L. caused by *Colletotricum piperi* Petch. *Science*. 16(2), 87–91.

Bhale, M. S., Chaurasia, R. K., and Nayak, M. L. (1985). Association of *Colletotrichum capsici* with *Xanthomonas campestris* pv. *betlicola* incident of leaf spot of betelvine (*Piper betle* L.). *Indian Phytopath.* 38, 565–566.

Bhale, M. S., Khare, M. N., Nayak, M. L., and Chaurasia, R. K. (1987). Diseases of betelvine (*Piper betle* L.) and their management. *Rev. Trop. Plant Pathol.* 4, 199–220.

Breed, R. S., Bergey, D. H., Parker Hitchens, A., and Murray, E. G. D. (1948). Bertchensgey's manual of determinative bacteriology. *American Society for Microbiology*, eBook. Geneva, N.Y.: Biotech, p.130.

Chattopadhyay S. B. and Maity S. (1967). *Diseases of betelvine and species*. New Delhi: ICAR.

Chattopadhyay, S. B. and Sengupta, S. K. (1955). A new disease of *Piper betle* in West Bengal. *Indian Phytopath.* 8, 105–111.

Chaurasia, J. P. (1994). Studies on the management of Betelvine-Phytophthora disease in Sagar division. *Ph.D. Thesis*, Dr. H. S. Gour University, Sagar. p.110.

Chowdhury, S. (1945). Diseases of *pan (Piper belle)* in Sylhet, Assam. VI. *Gloeosporium* leaf spot. *Proc. Indian Acad. Sci.* B. 22, 189–190.

Dasgupta, B., Dutta, P., and Das, S. (2011). Biological control of foot rot of Betelvine (*Piper betle* L.) caused by *Phytophthora parasitica* Dastur. *J. Plant Protection Sci.* 3(1), 15–19.

Dasgupta, B. and Maiti, S. (2008). Research on Betel vine diseases under AINP on betel vine. *Proceedings of National Seminar on "Piperaceae Harnessing Agro-technologies for Accelerated Production of Economically Important Piper Species,"* November 21–22, 2008, Indian Institute of Spices Research, Calicut, Kerala, pp. 270–279.

Dasgupta, B. and Sen, C. (1999). Assessment of *Phytophthora* root rot of betelvine and its management using chemicals. *J. Mycol. Pl. Path.* 29, 91–95.

Dasgupta, B., Roy, J. K., and Sen, C. (2000). Two major fungal diseases of betelvine. In: M. K. Dasgupta (ed.), *Diseases of Plantation Crops, Spices, Betelvine and Mulberry*. Bolpur: Palli Shiksha Bhavana, Visva Bharati, pp. 133–137.

Dastur, J. F. (1927). A short note on the foot rot diseases of pan in central provinces. *Agric. J. India*. 22, 105–108.

Deshpande, A. L. (1986). Preliminary report on mosaic disease of betelvine. Presented at the *3rd Workshop on Betelvine Diseases. Bapatla Problems and Progress in Betelvine Pathology in Madhya Pradesh*, pp. 21–23.

Goswami, B. K., Kader, K. A., Rahman, M. L., Islam, M. R., and Malaker, P. K. (2002). Development of leaf spot of betelvine caused by *Colletotrichum capsici. Bangladesh J. Plant Pathol.* 18(1&2), 39–42.

Hector, G. P. (1924). Annual Report Department of Agriculture. Bengal. *1923–1924*.

Hector, G. P. (1927). *Annual Report of Department of Agriculture, Bengal 1925-1926.*

Jhamaria, S. L. and Daftari, L. N. (1970a). A new disease of *Piper betle. Indian Phytopath.* 23, 130.

Latorre, B. (2004). *Enfermedades de las plantas cultivadas*. Santiago: Ediciones Universidad Católica de Chile. p. 638.

Magdum, S. G., Shirke, M. S., Kamble, B. M., Salunkhe, S. M., and Tambe, B. N. (2009). Integrated management of major diseases of betelvine. *Adv. Plant Sci.* 22(1), 35–36.

Maiti, S. (1994). Diseases of betel vine, In: Chadha, K. L., Rethinam, P. (eds.), *Advances in Horticulture Vol. 10, Plantation Crops and Spices Crops—Part 2.* New Delhi: Malhotra Publishing House.

Maiti, S. and Sen, C. (1977). Leaf and foot rot of *Piper betle* caused by *Phytophthora palmivora. Indian Phytopath.* 30: 438-39.

Maiti, S. and Sen, C. (1979a). Leaf rot and foot rot of *Piper betel* caused by *Phytophthora palmivora. Indian Phytopath.* 30, 438–439.

Maiti, S. and Sen, C. (1979b). Fungal diseases of betel vine. *Proc. Acad. Natl Sci.* 25, 150–157.

Maiti, S. and Sen, C. (1982). Incidence of major diseases of betel vine in relation to weather. *Indian Phytopathol.* 35(1), 14–17.

Mahesha, L., Nadugala, N. S., and Amarasinghe, B. H. R. R. (2009). Diversity among different isolates of *Xanthomonas campestris* pv. *betlicola* on the basis of phenotypic and virulence characteristics. *J. Natl Sci. Found. Sri Lanka*. 37(1), 77–80.

Mathew, J., Chirian, M. and Abraham, K. (1978). Bacterial leaf spot of betel vine (*Piper betle* L.) incited by *Xanthomonas betlicola* Patel et al., in Kerala. *Curr. Sci*. 16, 592–593.

Mathur, R. S. and Sinha, R. P. (1959). Control of foot rot of pan, *Piper betle* L. in Uttar Pradesh. *Pl. Prot. Bull. New Delhi*. 8, 17.

McRae, W. (1928). Reopen of the imperial mycologist, 1926-27. *Sci. Rep. Agric. Res. Inst. Pusa.*, pp. 45–55.

McRae, W. (1930). Reopen of the imperial mycologist, 1928-29. *Sci. Rep. Agric. Res. Inst. Pusa.*, pp. 51–66.

McRae, W. (1934). Foot rot diseases of *Piper belle* in Bengal. *Indian J. Agric. Sci*. 4, 585–617.

Mitra, M. (1930). India: New diseases reported during the year. *Int. Bull. Pl. Prot*. 4(7), 103–104.

Narasimhan, K. J. (1933). *Annual Report of the Mycology Section for the Year 1931-32*. Mysore: Department of Agriculture, pp. 32–35.

Nema, A. G. and Nayak, M. L. (1975). Powdery mildew in betel vine and its control. *Pesticides*. 9(3), 30–31.

Patel, M. K., Kulkarny, Y. S., and Dhande, G. W. (1951). Bacterial leaf spot of castor. *Curr. Sci*. 20, 106.

Raghunathan, C. (1926). Bacterial leaf spot of betel leaf *Ceylon Dep. Agric*. 39, 2

Raghunathan, C. (1928). Bacterial leafspot of betel. *Ann. Rep. bot. Gdn Peradeniya* 11, 51–56.

Rao, A. P., Karunakar Babu, M., and Rao, D. V. S. (1986). *Annual Progress Report for the Year 1985*. Ponnur: All India Coordinated Research Project on Betelvine Chintalapudi. p. 5.

Rao, N. N. R. (1993). Assessment of loss in leaf yield due to powdery mildew disease in Betelvine. *Indian J. Plant Protect*. 21(1), 94.

Rao, Z. V. (1987). Investigations into mosaic disease of Betelvine (*Piper betle* L.). *MSc (Agric.) Thesis*. Andhra Pradesh Agricultural University.

Roy, T. C. (1948). Anthracnose disease of *Piper betle* caused by *Colletotricum dasturi*. *Roy. J. India Bot. Soc*. 27, 96–100.

Shaw, F. J. F. and Ajrekar. S. L. (1915). The genus *Rhizoctonia* in India. *Mem. Dep. Agric. India (Bot. Ser)*. 7, 177–194.

Singh, B. P. and Chand, J. N. (1971). Studies on the diseases of pan (*Piper betle*, L.) in Jabalpur (Madhya Pradesh) IV. A new fungal-bacterial complex. *Sci. Cult*. 37(7), 344.

Singh, B. P., Joshi, L. K. (1971). Studies on the diseases of betel vine. *Indian J. Mycol. Plant Pathol*. 1(2), 150–151.

Singh, B. P. and Joshi, L. K. (1972). Studies on the diseases of *Piper betle* in Jabalpur (Madhya Pradesh). Four new world records of fusaria. *Sci. Cult*. 38, 285.

Singh, S. J. (1986). Coordinated research project on betelvine: Project Coordinators Report on Progress of Work Done for the year 1985. Bangalore: Indian Institute of Horticultural Research, pp. 10.

Singh, S. J. and Rao, N. N. R. (1988). Betelvine mosaic: A new virus disease. *Curr. Sci*. 57, 1024–1025.

Stevenson, A. (1926). *Foreign Plant Diseases*. Washington, D.C.: U.S. Department of Agriculture.

Uppal, B. N. and Kamat, M. N. (1938). Powdery mildew of betel vine. *Curr. Sci*. 12, 611.

CHAPTER 3

MAJOR DISEASES OF COCOA OR CHOCOLATE (*THEOBROMA CACAO* L.) AND THEIR MANAGEMENT

RAJU J.[1], JAYALAKSHMI K.[2*], PRITI S. SONAVANE[3], and RAGHU S.[4]

[1]*Plant Quarantine Station, Ministry of Agriculture and Farmers Welfare, Government of India, Mangalore, Karnataka 575011, India*

[2]*AINRP (Tobacco), ZAHRS, University of Agricultural and Horticultural Sciences, Shimoga, Karnataka 577201, India*

[3]*Central Horticultural Experiment Station, ICAR-IIHR, Chettalli, Kodagu, Karnataka 571248, India*

[4]*Crop Protection Division, National Rice Research Institute, Cuttack, Odisha 753006, India*

[*]*Corresponding author. E-mail: jayalakshmipat@gmail.com*

ABSTRACT

Cocoa (*Theobroma cacao* L.) belongs to the family Malvaceae. It is an evergreen tree grown for its seeds (beans). Cocoa may also be referred to as cacao tree, koko, and Kacao. Cocoa seeds are majorly used in the manufacture of chocolate. The present chapter discusses about the diseases that are known to cause economic loss of cocoa.

3.1 SEEDLING DIEBACK DISEASE

Symptoms:

1. One-to-four-month-old seedlings are more prone to disease.
2. Initial infection starts from the tip of the stem or from the collar region.

3. Characteristic symptoms of this disease are defoliation and dieback of seedling.
4. Mainly the infection starts from the tip of the stem and proceeds downward as dark brown to black water-soaked linear lesions.
5. Later in the subsequent growth, the lesions also extend to the leaves through the petioles, resulting in wilting and subsequent defoliation of the seedlings.

Causal Organism: *Phytophthora palmivora.*

Disease Cycle:

The primary source of inoculums is Oospores present in the soil, and the secondary spread takes place through zoospores and sporangia (Gregory, 1974).

Epidemiology:

Continuous heavy rainfall with intermittent bright sunshine hours, low temperature (20 °C–23 °C), and high relative humidity (>90%) are factors congenial for disease development (Griffin et al., 1981).

Management:

1. Complete removal and destruction of infected seedlings.
2. Adequate drainage should be provided.
3. *Trichoderma viride* applied along with farm yard manure.
4. Just before the onset of monsoon soil drenching with 1% Bordeaux mixture or 0.3% copper oxychloride, and thereafter at frequent intervals.

3.2 BLACK POD DISEASE

It occurs in the rainy season (June–September). Plants of all ages are susceptible to pods. The black pod disease was first reported in the world from West Indies in 1897, and now it is widespread in cocoa-growing countries. In India, the disease was first reported in 1965. This disease is also reported in Karnataka, Kerala, and Tamil Nadu.

Symptoms:

The infection appears as one or more brown, small circular, lesions anywhere on the pod surface. The lesions increase rapidly and cover the whole surface of the pod. Ultimately, the whole pod and the beans are infected by the pathogen, and the pod turns black in color.

Causal Organism: *Phytophthora palmivora* (*P. palmivora*) (Figure 3.1) and *P. megakarya* (*P. megakarya*) (Figure 3.2).

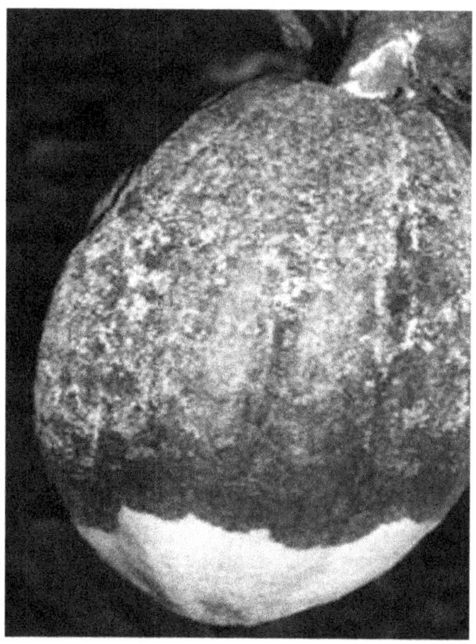

FIGURE 3.1 Black pod rot: *Phytophthora palmivora*

FIGURE 3.2 Black pod rot: *Phytophthora megakarya*

Disease Cycle:

Oospores present in the soil and infected pods act as the primary source of inoculum, and the secondary spread takes place through zoospores and sporangia.

P. palmivora-infected single mummified pod is able to produce up to 4 million sporangia under humid conditions which can be disseminated by rain, flying insects, ants, rodents, bats, and contaminated pruning material. In the case of *P. megakarya*, sporulation is usually more abundant. The soil-borne phase of the *P. megakarya* disease cycle causes root infection, maintaining a reservoir of inoculum that releases zoospores into the soil surface water. *P. megakarya* does not survive in mummified pods, but it can survive in infected debris for at least 18 months, while *P. palmivora* survives less than 10 months in the soil (Gregory, 1974).

Epidemiology:

Heavy rainfall with intermittent bright sunshine hours, low temperature (20 °C–23 °C) and high relative humidity (>90%) are factors congenial for disease development.

Management:

1. Roughing of all infected pods.
2. Adequate spacing should be provided. Since high humidity and low temperature are favorable for disease development, cocoa should not be planted very closely. Pruning of cocoa trees is also very essential to reduce the shade effect.
3. During the onset of monsoon spraying of 1% Bordeaux mixture.
4. Weekly removal of infected pods and spraying Bordeaux mixture (1%) at frequent intervals depending on the severity of the disease will give good control of the disease.

3.3 STEM CANKER

Symptoms:

1. It affects both seedlings and mature plants.
2. It is difficult to detect the canker in early stages because the outer bark appears quite normal.
3. It also affects the main trunk and fan branches.

4. Conspicuous symptoms with grayish brown water-soaked lesions with a broad dark-brown to black margin on the bark.
5. Liquid oozes of reddish brown color out from the cracks forming a rusty deposit.
6. The canker at the base of the stem appears as irregular, dark-brown, water-soaked lesion with oozing of reddish-brown liquid.
7. Reddish brown discoloration observed on the internal bark beneath the outer lesion appears as lesions in the internal bark coalesce, leading to extensive rotting.
8. The pathogen invades from the cortical tissues into vascular tissues reaching wood.
9. Grayish brown to black discoloration with black streaks appears on the wood.
10. Dieback symptoms occur when the cankers girdle the main stem or branches, leading to the collapse of the entire tree.
11. The canker from the pods spreads to the peduncle and then to the cushion and bark causing canker (Figure 3.3).

FIGURE 3.3 Stem canker.

Causal Organism: *Phytophthora palmivora.*

Management:

1. Follow similar management practices recommended for the black pod rot disease.
2. Diseased bark along with healthy tissues should be removed and the Bordeaux paste should be applied.

3.4 CHARCOAL POD ROT

Symptoms:

1. Disease occurs throughout the year, and it becomes severe during summer months.
2. Pods of all the stages are susceptible.
3. Dark-brown to black-colored spot appears on the pod.
4. At later stages, the affected spots turn to black and remain on the tree as a mummified fruit.
5. Finally, internal tissues get rotten and the affected beans turn black (Mbenoun et al., 2008).

Causal Organism: *Lasiodiplodia theobromae* (*Botryodiplodia theobromae*).

Disease Cycle:

The fungus overwinters as pycnidia on the outside of diseased wood. The pycnidia produces and releases two-celled, dark-brown, striated conidia. The conidia are then dispersed by wind and rain splash (Kannan et al., 2009).

Management:

1. Removal and destruction of infected pods.
2. Spraying with 1% Bordeaux mixture is recommended to manage the disease.

3.5 DIEBACK

Dieback of cocoa was first observed in Hunsur Taluk of Mysore district, Karnataka, during September 2000. All growth stages of the tree are susceptible to disease it resides throughout the year, but the mortality was most severe from March to May.

Symptoms:

1. Affected trees show sudden yellowing and browning of leaves, followed by the rapid dieback of branches from the tips (Figure 3.4).
2. Infected twigs have brownish, short vertical streaks in the vascular tissues (Mohali et al., 2005).
3. Affected trees continue to decline for several days and soon they die, although new flushes still develop at the collar regions during the monsoon (Ko et al., 2004).
4. Pods do not develop in diseased trees, but if formed earlier, become mummified.

FIGURE 3.4 Seedling dieback.

Causal Organism: *Lasiodiplodia theobromae.*

Disease Cycle:

Overwintering of fungi occurs through pycnidia on the outside of diseased wood. Two-celled, dark-brown, striated conidia are released from the pycnidium. Conidia are dispersed by wind and rain splash from one part of the vine to another, spreading the fungi to other vines. When conidia land on freshly cut or damaged wood, the disease develops (Khanzada et al., 2004). The conidia then germinate on the tissue of the wood and start causing damage to the vascular system of the plants. Necrosis and dieback of the wood are noticed after the complete damage of the vascular system. In some cases, pseudothecia form on the outside of the tissue and produce ascospores which are then dispersed like conidia (Keane et al., 2009).

Epidemiology:

Relative humidity above 80% and temperature of 25 °C–31 °C favor the development of disease.

Management:

1. Diseased twigs have to be pruned and spray with Bordeaux mixture (1%) copper oxychloride (0.3%) on infected trees. Diseased twigs are removed 2–3 inches below the affected portion (Pedraza et al., 2013).
2. In small plants, pruning of twigs is followed by pasting of copper oxychloride or bordo paste.

3.6 VASCULAR STREAK DIEBACK

During the 1960s in Papua New Guinea, vascular streak dieback (VSD) (Figure 3.5) was first distinguished. It was recognized as the most destructive disease in mature plantations of cocoa in Papua New Guinea, Malaysia, Indonesia, and Philippines (Punithalingam, 1976). At present in India, VSD has been observed in Thiruvananthapuram, Kottayam, Kozhikode, and Wayanad districts of Kerala state. The disease causes dieback of canopy and can kill young bushes. During the establishment phase, VSD is more savior (Purwantara et al., 1977). Mature cocoa trees can usually survive the attack of fungi, but varying degrees of yield losses may be expected depending on the severity of the disease and the susceptibility of the cocoa.

Symptoms:

1. Yellowing of the growing tip of one or two leaves on the second or third flush.
2. Leaves turn yellow and diseased leaves fall within a few days.
3. Similar symptoms are also observed on other leaves.
4. Rapid death of the axillary bud is noticed. At later stages, dieback symptoms appear on these branches.
5. A characteristic brown streaking of the woody tissue is observed when the infected shoot is split lengthwise.
6. Fruiting bodies of the fungus appear as a white crust around the leaf scars in wet weather. Leaves in the diseased seedling or branch show interveinal necrosis similar to symptoms of calcium deficiency.

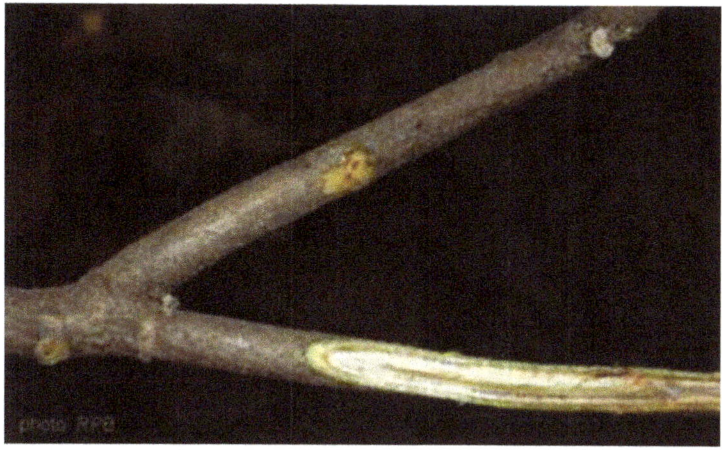

FIGURE 3.5 Vascular streak dieback.

Causal Organism: *Oncobasidium theobromae*.

Epidemiology:

Mainly during nighttime, the formation and forcible discharge of basidiospores occur. Spores are dispersed by wind, although effective spore dispersal is limited by high humidity. The basidiocarps are developed when leaf fall occurs during wet weather. They have a short lifespan, and when the basidiocarps are sufficiently moist, they release spores only at night (Punithalingam, 1976).

Management:

1. Inoculum levels are reduced when pruning of the diseased material is about 30 cm below the discolored xylem which prevents further expansion of infection.
2. Disease-free seedlings will be raised by planting material under plastic cover and away from infected plantations.
3. Disposal of diseased branches and regular pruning of chupons on the trunk.
4. Cocoa nurseries should be raised away from the diseased trees.
5. Overcrowding of trees and thick shade should be avoided.
6. Spraying of 0.1% benomyl or propiconazole reduces the infection.
7. Kerala Agriculture University has developed some VSD resistant and high-yielding varieties (CCRP-1 to CCRP-7).

3.7 FROSTY POD

Symptoms:

1. Actively growing pod tissues are affected by fungus, especially young pods.
2. The time from infection to the development of symptoms is about 1–3 months.
3. The most outstanding symptom is the white fungal mat on the pod surface.
4. Spots on the surface of immature pods and the spots turning brown and rapidly enlarging to cover the entire pod surface (Figure 3.6).

FIGURE 3.6 Frosty pod rot.

Causal Organism: Basidiomycetes = *Moniliophthora roreri*

Disease Cycle:

The large number of spores is produced, and the genetic variability endows the fungus with considerable adaptability. The dry, powdery form of spores dislodged by water, wind, or physical disturbance of the pod. Disease incidence varies with cultivar, age of the pods, and rainfall. Generally, the greatest production of spores is when rainfall is very high.

Epidemiology:

Temperature of 22 °C–26 °C and a relative humidity of 93% favor the disease development.

Management:

1. Removal and destruction of infected pods.
2. Planting of cocoa varieties that produce pods during the dry season avoids the disease development.
3. Spraying of copper containing fungicides (1% Bordeaux mixture).

3.8 COLLETOTRICHUM POD ROT/CHERELLE ROT

Symptoms:

1. A large number of young, developing pods known as "cherelle" of 2–3 months old dry up and remain on the tree as a mummified fruit, commonly referred to as "cherelle wilt/rot." It is noticed especially during February–May.
2. The symptom mostly starts from the stalk end, particularly at the point of attachment of the stalk to the pod, and proceeds toward the tip of the pod as dark-brown sunken lesions with a diffused yellow halo (Nambiar et al., 1972).
3. The infection also extends to the stalk and reaches the cushion but does not spread further in the cushion.
4. As the infection progresses, the internal tissue of the pod becomes discolored.
5. The infection may start from anywhere on the pod surface, other than the stalk region, as a dark-brown sunken lesion (Nambiar et al., 1972).

6. Such lesions coalesce and form bigger lesions with salmon/dark-brown-colored fruiting bodies of the fungus.
7. Ultimately, the cacao pod turns dark brown to black color and remains mummified on the tree trunks.
8. At this stage, these pods can be easily confused for pods affected by the cherelle wilt, which is a physiological phenomenon.

Causal Organism: *Colletotrichum gloeosporioides.*

Disease Cycle:

Survives as a dormant mycelium in the infected cherelles or pods and spread through wind-borne and rain-splashed conidia.

Epidemiology:

Warm wet weather with 25 °C–30 °C temperature and 80% relative humidity favor disease development.

Management:

1. Collect and destroy all infected and dried cherelles.
2. Remove and destroy by burning the piles of pod husks lying in the garden.
3. Regularly harvest the ripe pods to ensure that recently infested pods are not completely lost.
4. Spray of 1% Bordeaux mixture as prophylactic measure or 0.3%–0.4% copper oxychloride, or 0.3%–0.4% mancozeb (at 40–45-day intervals, especially on pods and flower cushions during summer months (Chowdappa et al., 1999).
5. In diseased gardens, spray the pod bunches twice with 0.1% hexaconazole or propiconazole or 0.1% carbendazim at 2–3-week intervals.

3.9 PINK DISEASE

Symptoms:

1. The first indication of the development of disease is oozing of brown liquid, usually from the fork region and also from the trunk and branches.
2. White, silky threads (mycelia) of the fungus appear and spread around the branch like a "cobweb."
3. Shoots wilt, leaves shed, and ultimately branches dry up.

4. As infection progresses, there is more bleeding, open wounds are formed, and trunk/branches become ringed and die.
5. Pinkish powdery coating appears on the irregular cracks as a characteristic symptom.
6. Below the point of infection numerous offshoots are formed.

Causal Organism: *Phanerochaete salmonicolor* (*Corticium salmonicolor*).

Disease Cycle:

Wind and rain help for the dispersal of spores. They need water to germinate, and then they infect through healthy bark.

Management:

1. Remove and destroy completely infected and dried shoots/branches.
2. Reduce overhead shade.
3. Pruning is practiced to improving aeration inside the garden. Prune smaller infected branches and swab the cut ends with the Bordeaux paste or 0.1% hexaconazole in rubber kote or coal tar (5 mL/kg).
4. As a preventive measure, 1% Bordeaux mixture is sprayed at regular intervals on forking regions and branches during the rainy season.

3.10 ROOT DISEASES

Symptoms:

1. Sudden premature yellowing and drying of slightly off-green leaves is a possible indication of root diseases (Figure 3.7).
2. Rhizomorphs are found firmly attached to the infected roots or ramifying into a network with soil encrustation around the root.
3. They are brown/red/black depending on the fungus associated.
4. The wood of a newly killed root has brown/black lines and is hard.
5. Generally, fruiting bodies of the fungi grow at the collars of diseased trees and on decaying exposed roots or stumps.
6. The disease can be identified based on the fruiting bodies or rhizomorphs grown on dead stumps or decayed roots.

Causal Organisms:

Brown root rot: *Phellinus noxius*.Red root rot: *Ganoderma* spp.
Black root rot: *Armillaria* spp. and *Rosellina arcuate*.

FIGURE 3.7 Root diseased trees.

Epidemiology:

Very low or high soil temperature, poor physical conditions of soil, moisture stress, inadequate shade, wounds on roots, etc. will aggravate the disease.

Management:

1. Dead/infected tree parts including roots are removed and destroyed. Treat soil with 2% CuSO4 (20 g/L) and replant after six months to one year.
2. Prune branches, regulate shade, and provide adequate spacing for improving aeration.
3. Apply organic matter @ 10–15 kg/pit and lime @ 1–2 kg/pit.
4. Isolate infected trees by digging trenches 1.2 m deep and 0.6 m wide.
5. Soil is drenched with 1% Bordeaux mixture or 0.2% copper oxychloride (2 g/L) or 0.2% hexaconazole or propiconazole (2 ml/L)/carbendazim (2 g/L).

3.11 WITCHES' BROOM

Causal Organism: *Moniliophthora perniciosa. It* is the causal agent for Witches' broom disease.

Symptoms:

1. Actively growing tissues (shoots, flowers, and pods) are attacked by the fungus causing cocoa trees to produce branches with no fruit and ineffective leaves.
2. Production of branches which do not produce pods. If pods are produced, they show distortion and green patches are present on pods that give the appearance of uneven ripening.

Causal Organism: *Moniliophthora perniciosa.*
Moniliophthora perniciosa is the causal agent for Witches' broom disease (Figure 3.8).

FIGURE 3.8 Witches' broom.

Disease Cycle:

The life cycle of the fungus is synchronized with the host phenology. Water is one of the most influential factors for the adequate reproduction of the fungus.

At night, basidiospores are released and are related to the level of humidity and favorable temperature. The spores are capable of being disseminated by water and convection currents and over long distances by wind.

Epidemiology:

Disease development is greatly influenced by humidity with emergence favored by temperature (20 °C–30 °C) and high humidity (>80%).

Management:

1. Removal and destruction of affected parts.
2. Spraying of 0.1% propiconazole or 0.2% carbendazim or 1% Bordeaux mixture.

3.12 COCOA SWOLLEN SHOOT VIRUS

Symptoms:

1. Swelling of leaves and shoots, chlorotic patches next to leaf veins.
2. Intense redness along the principal veins and between the secondary veins of the young cacao leaf (Dzahini-Obiatey and Adu Ampomah, 2010).
3. Chlorotic spots or flecks on leaves mottled.
4. Smooth pods with reduced beans; mottled coloration on pods.
5. Swellings are developed on stems at nodes or internodes and at shoot tips.
6. Progressive defoliation may occur ultimately leading to the death of the tree.

Causal Organism: Cocoa swollen shoot virus (CSSV). The virus belongs to badnavirus within the family Caulimoviridae.

Transmission:

Mealybugs are the primary transmitting agents of CSSV (End et al., 2017). These mealybugs develop a mutualistic relationship with ants which provide protection in return for sugar exudates (Figure 3.11). *Planococcoides njalensis* and *Planococcus citri* are the most important mealybugs among others, and they act as a vector for the transmission of virus in a semipersistent manner (Dzahini-Obiatey et al., 2010).

FIGURE 3.9 Symptom of cocoa swollen virus on shoot.

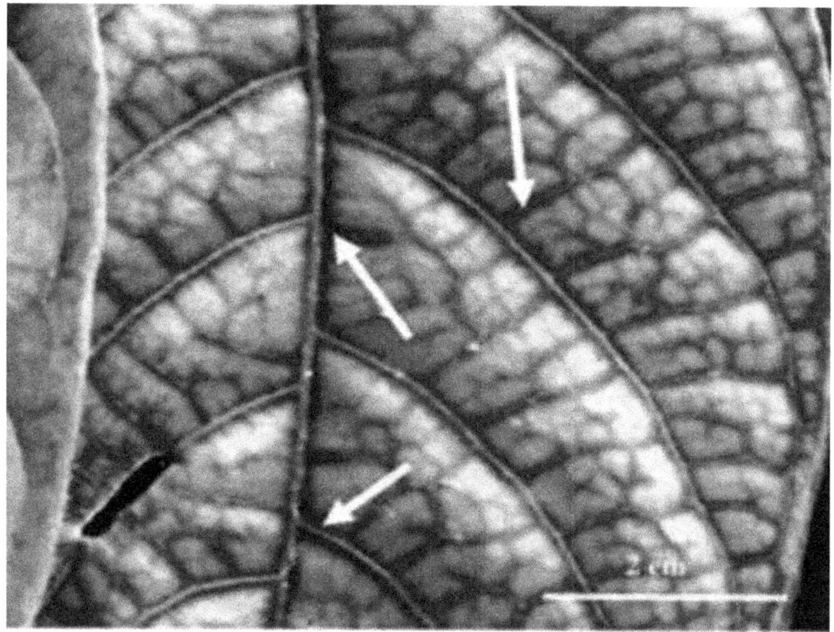

FIGURE 3.10 Symptom of cocoa swollen virus on leaf veins.

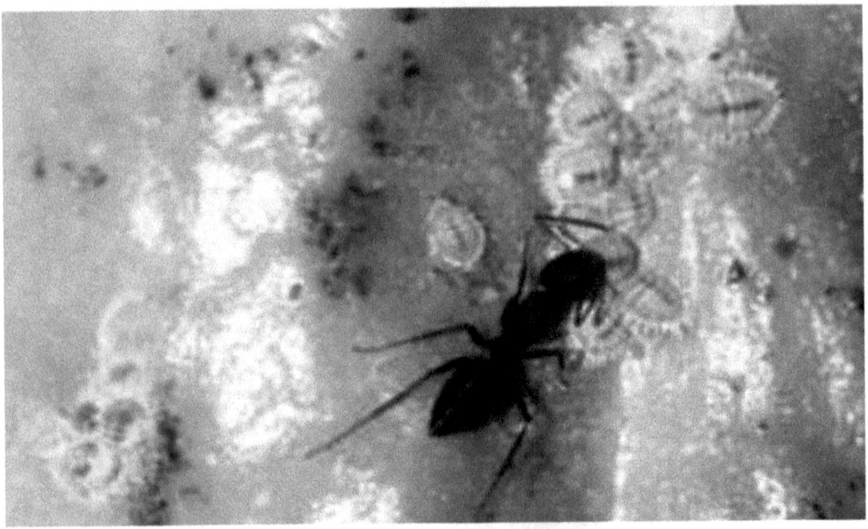

FIGURE 3.11 Cocoa swollen shoot virus vector: *Planococcoides* sp. tended by a black ant.

Management:

1. Infected trees parts and those surrounding them should be removed and destroyed.
2. Using disease free planting material cultivation
3. Spraying of systemic insecticides to control mealybugs.
4. Growing of barrier crops such as oil palm, coffee, cola, and citrus to avoid the movement of vectors.

3.13 CHERELLE WILT

Symptoms:

1. This is a "physiological" condition wherein flowers are produced, they are pollinated by midges, and cherelles—the young pods—develop, but many die. It is a natural event. It is a fruit-thinning process, probably controlled by growth hormones produced by plants. Generally, it does not affect the yield of the crop. The cocoa tree allows as many cherelles to develop into mature pods as there are nutrients to support them.
2. The shriveling and mummifying of some young fruits in all coca gardens is a common phenomenon.

3. In the early stages of crop growth, fruits lose their luster and in 4–7 days and shrivel. The fruits may wilt but do not abscise.

Causal Organism: Many factors are involved in the causation of cherelle wilt (Figure 3.12). The most important factors are insects, fungi (*Phytophthora palmivora*), nutrient competition, overproduction, etc.

FIGURE 3.12 Cherelle wilt.

Management:

1. Application of fertilizers and mulches should be practiced. The greater the availability of nutrients, the fewer cherelles wilt.
2. The correct balance is needed between sunlight for the manufacture of plant foods and sunscald that damages foliage.
3. Closer planting of cacao plants should be avoided.
4. Severity of cherelle wilt will increase if trees are too crowded because of competition for essential nutrients, water, light, and others.
5. Keep soils moist, either with the help of drains or mulches, as cherelle wilt is severe in soils that are either too dry or too wet.
6. Control fungi and insects. This is probably secondary in terms of importance, unless there is an epidemic of, for example, black pod disease.

KEYWORDS

- **coco disease**
- **etiology**
- **symptomatology**
- **management**

REFERENCES

Chowdappa, P., Saraswathy, N., and Iyer, R. (1999). Disease management strategies in arecanui and cocoa. Extension Publication, Director, CPCRI, Kasaragod, Kerala.

Dzahini-Obiatey, H. and Adu Ampomah, Y. (2010). Cocoa swollen shoot virus: Genus *Badnavirus*. In: End, M. J. (ed.), *Technical guidelines for the safe movement of cacao germplasm* (Revised from *FAO/IPGRI Technical Guidelines*, No. 20). Montpellier, France: Bioversity International.

Dzahini-Obiatey, H., Domfeh, O., and Amoah, F. M. (2010). Review: Over seventy years of a viral disease of cocoa in Ghana: From researchers' perspective. *African Journal of Agricultural Research*, 5(7), 476–485.

End, M. J., Daymond, A. J., and Hadley, P. (eds.). (2017). Description of pests of cacao. *Technical guidelines for the safe movement of cacao germplasm* (pp. 20-22) (Revised from the *FAO/IPGRI Technical Guidelines*, No. 20 (Third Update, October 2017)) Global Cacao Genetic Resources Network (CacaoNet). Montpellier, France: Bioversity International.

Gregory, P. H. (ed.). (1974). *Phytophthora disease of cocoa*. London: Longman.

Griffin, M. J., Idowu, A. C., Maddison, A. C., Taylor, B., and Ward, M. R. (1981). Sources of infection. In: Gregory, P. H. and Maddison, A. C. (eds.), *Epidemiology of* Phytophthora *on cocoa in Nigeria*. Kew, England: Commonwealth Mycological Institute.

Kannan, C., Karthik, M., and Priya, K. (2009). *Lasiodiplodia theobromae* causes a damaging dieback of cocoa in India. *New Disease Reports*, 19, 63.

Keane, P. (2009). New symptoms of vascular-streak dieback of cocoa in Southeast Asia e possible causes, studies required and control. *COPAL International Cocoa Research Conference*, No.16, Bali, November 16–21.

Khanzada, M. A., Lodhi, A. M., Shahzad, S. (2004). Mango dieback and gummosis in Sindh Pakistan caused by *Lasiodiplodia theobromae*. Plant Health Progress. On-line [http://www.plantmanagementnetwork.org].

Ko, W. H., Wang, I. T., and Ann, P. J. (2004). *Lasiodiplodia theobromae* as a causal agent of kumquat dieback in Taiwan. *Plant Disease*, 88, 1383.

Mbenoun, M., Momo Zeutsa, E. H., Samuels, G., Nsouga Amougou, F., and Nyasse, S. (2008). Dieback due to *Lasiodiplodia theobromae*: A new constraint to cocoa production in Cameroon. *Plant Pathology*, 57, 381.

Mohali, S., Burgess, T. I., and Wingfield, M. J. (2005). Diversity and host association of the tropical tree endophyte *Lasiodiplodia theobromae* revealed using simple sequence repeat markers. *Forest Pathology*, 35, 385–396.

Nambiar, K. K. N. and Nair, R. R. (1972). Charcoal pod rot of cacao. *Indian Phytopathology*, 25, 595–597.

Pedraza, J. M. T., Aguilera, J. A. M., Díaz, C. N., Ortiz, D. T., Monter, Á. V., Mir, y S. G. L. (2013). Control of *Lasiodiplodia theobromae*, the causal agent of dieback of sapote mamey [*Pouteria sapota* (Jacq.) H. E. Moore and stearn] grafts in México. *Revista Fitotecnia Mexicana*, 36(3), 233–238.

Punithalingam, E. (1976). *CMI descriptions of pathogenic fungi and bacteria*, No. 519. Wallingford, UK: CAB International.

Purwantara, A., Iswanto, A., Sukamto, S., McMahon, P., Purung, H., Lambert, S., Guest, D., and Varma, U. and Bilgrami, K. S. (1977). New host records of *Botyrodiplodia theobromae*. *Indian Phytopathology*, 30, 579.

CHAPTER 4

COCONUT (*COCOS NUCIFERA* LINN.) DISEASES AND MANAGEMENT STRATEGIES

R. RAMJEGATHESH, L. RAJENDRAN*, G. KARTHIKEYAN, and T. RAGUCHANDER

Department of Plant Pathology, Tamil Nadu Agriculture University, Coimbatore 641003, Tamil Nadu, India

**Corresponding author. E-mail: rucklingraja@gmail.com.*

ABSTRACT

The coconut palm (*Cocos nucifera* Linn.) is a major plantation as well as oilseed crop and much adapted for cultivation in various climatic zones, varying from islands, seashores, plains, to hills. This crop is predominantly cultivated in the southern states of India, namely Tamil Nadu, Kerala, Karnataka, and Andhra Pradesh. As per the India estimates for the year 2014–2015, the cultivated area and production in the country is 1.98 million hectares and 20439.61 million nuts, respectively. The four southern states of Kerala, Karnataka, Tamil Nadu, and Andhra Pradesh accounted for 87.86% the area and 90.11% of the production. This perennial palm is often infected by plant pathogens due to moisture stress and climatic factors (Henry Louis, 2002). More than 50 diseases have been reported in different parts of the world in coconut in spite of its hardy nature, and 173 fungi were associated with coconut (Anon, 1979) and 35 fungi were associated with coconut leaves (Brown, 1973) in India. Besides these, bacteria, phytoplasma, virus, and nematodes were also reported to cause disease (Nambiar, 1994; Srinivasulu et al., 2001). The following diseases are serious constraints to the production and productivity.

4.1 LEAF BLIGHT

Introduction/Economic Importance:

The disease was first noticed in 1994 in South India, especially in Coimbatore district of Tamil Nadu. Nowadays, the disease is spreading very quickly in neighboring districts. The disease was noticed in young seedling to older plants of coconut and oil palms. Especially, adult plants of coconut and oil palms are highly susceptible in nature. Normally, the disease occurred in gardens where more nitrogenous fertilizers were applied.

Symptoms:

This pathogen infects leaflets, fronds, and nuts, and the disease occurred in the adult fronds containing leaflets in the lower (3-10) whorls. Heavily infected coconut palms show delayed flowering than healthy palms (Abad and Blancaver, 1975), and the incidence is severe in older leaves and the younger leaves completely free from disease. Further, the infected leaflets start drying from the tip downward, and the drying leaflet margins are dark gray to brown with wavy to undulate, and the lesion spreads throughout the leaflets, and then the fronds exhibit a charred or burnt appearance. The disease reduces the vigor in the coconut seedlings by lesser leaf production and stunted growth. The fungus that enters into the kernel through the mesocarp results in the decay of the endosperm, and the nuts were desiccated, shrunk, deformed, and dropped prematurely. The nut yield loss extends up to 10%-25% in severe cases (Lakshmanan and Jegadeesan, 2004).

Casual Organism: *Lasiodiplodia* (*Botryodiplodia*) *theobromae* (Pat.).

Pathogen:

The disease caused by *Lasiodiplodia* (*Botryodiplodia*) *theobromae* (Pat.) Griffon and Maubl. is an emerging, serious problem in southern part of India, especially in Tamil Nadu. The fungal colonies were gray to black, fluffy with abundant aerial mycelium, reverse fuscous to black, and produced pycnidia that are stromatic, ostiolate, and frequently setose in nature. Conidiophores are hyaline, simple, sometimes single septate, and rarely branched, arising from the inner layers of cells lining the pycnidial cavity. Conidia are initially unicellular, hyaline, subovoid to ellipsoid-oblong, thick walled, and base truncate. Mature conidia are one-septate, cinnamon to fawn, and often longitudinally striate.

Disease Cycle:

The pathogen overwinters as pycnidia on the outside of diseased wood and releases two-celled, dark-brown, striated conidia, which are then dispersed by wind and rain splash, spreading to other plants or branch to another. When conidia land on the freshly cut or damaged wood, the disease starts and causes damage to the vascular system. Cankers begin to form around the initial infection, and eventually complete damage leads to necrosis and dieback.

Epidemiology:

The infection was noticed throughout the year. It was maximum in summer and low in cooling/winter months. The infection was favored by high temperature and humidity. The conidia and the resting structures on the infected portion of the leaves served as inoculums for further spread.

Management:

1. Removal and burning of severely infected leaves.
2. Spraying of Bordeaux mixture (1%) or copper oxychloride (0.25%) along with the sticking agent (1 m/L) for two times at a 15-day interval during summer months.
3. Carbendazim (2 g) or hexaconazole (2 mL) in 100 mL water was used for root feeding for three times at the interval of three months.
4. Soil application of *Pseudomonas fluorescens* (Pf1) @ 200 g along with 50 kg of farm yard manure (FYM) or 5 kg of neem cake/palm/year (Meena et al., 2008; Johnson et al., 2014), and the application of an additional quantity of 1.5 kg of Murite of Potash.

4.2 GRAY LEAF SPOT

Introduction/Economic Importance:

This disease occurs on all the coconut-growing areas of tropical regions and was first reported from British Guyana in 1931 (Menon and Pandalai, 1958). It is usually a minor incidence but can be severe under crowded or wet condition and is favored by the insect damage. It is commonly apparent that the incidence points out the poor nutritional status.

Symptoms:

The disease is characterized by the appearance of minute yellow spots surrounded by a grayish margin on the outer whorl of leaves. This spot may be oval in shape, and gradually the center of the spots turns grayish white with a dark-brown margin with a yellow halo. Many spots are coalescing into irregular gray necrotic patches, and in severe cases, complete drying and shriveling of the leaf blade occur, providing a blighted or burnt appearance. In addition, a large number of minute globose or ovoid black acervuli appear on the upper surface of infected leaves (Uchida, 2004).

Casual Organism: *Pestalotiopsis (Pestalotia) palmarum.*

Pathogen:

The fungus *Pestalotiopsis (Pestalotia) palmarum* causes the disease and produces acervuli. These are cushion shaped, black in color, subepidermal, and break open to expose conidia and setae (black sterile structure). Conidiophores are short, simple, hyaline, and they form conidia at the tip. Conidia are five-celled; the middle three cells are dark, and the end cells are hyaline with 35 slender, elongated appendages at the apex (Espinoza et al., 2008).

Disease Cycle:

The *Pestalotiopsis* infection occurs (inoculum) probably by means of the conidia or fragmented spores (Espinoza et al., 2008). These inocula may survive in harsh weather conditions and may cause primary infections, whereas the secondary inoculum is produced on the diseased tissue and increase the severity of the disease. The sources of the inoculum are splashed water droplets (Hopkins and McQuilken, 1997; Elliott et al., 2004), flowers (Pandey, 1990), crop debris, disease stock plants, used growing media, soil, and contaminated nursery tools (McQuilken and Hopkins, 2004), wild plantations (Keith et al., 2006) and spores in the air (Xu et al., 1999).

Epidemiology:

This pathogen is highly weak in nature, and the black pycnidia appear as black minute specks on the upper surface of the leaf. The higher conidial germination was 93% on the upper surface than the lower surface (Lingaraju et al., 1987). Disease is highly favored by soils with potash deficiency, continuous rainy weather for 4-5 days, poorly drained soils, and strong wind.

Management:

1. Severely affected leaves should be removed and burnt.
2. Spraying of Bordeaux mixture @ 1% or copper oxychloride (0.25%) added with the sticking agent (1 mL/L) twice at a 30 day-interval during summer months.
3. Hexaconazole @ 2 mL or carbendazim 2 g in 100 mL water root feeding for three times at a three-month interval (Rahman et al., 2013).
4. An excess of 1.5 kg potash needs to be applied along with the recommended dose of fertilizers (Polomer and Bentonio, 1982).
5. Soil application of *Pseudomonas fluorescens* (Pf1) @ 200 g with 50 kg of well decomposed FYM or 5 kg neem cake around the basin of the infected palm.

4.3 LEAF ROT

Introduction/Economic Importance:

Leaf rot, a disease of fungal complex, often occurs over root (wilt)-affected palms (Srinivasan, 1991), and the infected palms slowly decline in yield. Leaf rot symptoms occur within 10-month-old seedling to mature palms. The first symptom to appear in seedlings is rotting of spindle leaves and then flaccidity, yellowing, and necrosis. Normally, farmers identify a palm as root (wilt) affected only when the leaf rot sets in. Farmers are not able to identify this disease at early stages when the palm showing flaccidity and marginal necrosis of leaflets.

Symptoms:

Small water-soaked lesions appear on the spindle leaves, which later enlarge, coalesce, and copiously lead to extensive rotting. In the advance stage of infection, the rotting spreads in the interior portion, the spindle leaf grows, the rotted portions dry up, turn dark brown to black, and rotted portion of the leaves break and are blown off in the wind. Initially, the rotting symptoms are developed on the distal end of the spindle leaf, so the spindle leaf is highly susceptible to this disease. In many cases, the rotten distal portions of the leaflet adhere to each other from top to bottom on both sides, thereby giving the appearance of fish bone. On drying, these portions of the leaflets completely drop off. Chlorophyll development of the mature tissues generally start rotting and slow down toward the base of the leaflets, so that the basal part of the rotting leaflets remains green in color and normal, giving a fan-like appearance. During summer months, the early stages of the disease occurring palms containing some spindles escape from rotting. Severity of the leaf rot symptom is more during winter, and during summer months, the leaflets show more of dry rot. All the ages of palms are highly susceptible, and the disease causes severe reduction in the photosynthetic area, which causes reduction in the yield. All the leaf rot affected palms are highly susceptible for red palm weevil attack (Chandra Mohan and Peter, 2008).

Casual Organism:

Helminthosporium halodes (*H. halodes*) (*Bipolaris halodes*), *Gleosporium* sp., *Curvularia* sp., *Gliocladium roseum* (*G. roseum*), *Pestalotia* sp., and *Fusarium* sp. were isolated from the diseased leaves (Menon and Nair, 1948; Anonymous, 1985). Radha et al. (1961) and Srinivasan and Gunasekaran (1993, 1996a, 1996b) isolated the following fungi: *Colletotrichum gloeosporioides* (*C. gloeosporioides*), *Exserohilum rostratum*, *Gliocladium vermoeseni*, *Cylindrocladium scoparaium*, *Fusarium solani*, *F. moniliforme* var. intermedium. *Thielaviopsis paradoxa* (*T. paradoxa*), *Rhizoctonia solani*, *Mortierella elongata*, *Curvularia* sp., *Acremonium* sp., *Thielavia microspor*, *X. terricola*, and *Chaetomium brasiliense.*

Epidemiology:

High relative humidity and low-temperature (20 °C-25 °C) conditions favor the disease development (Menon and Nair, 1951). *H. halodes*, *Gleosporium* sp., and *G. roseum* causing leaf rot infection were found to be growing under high-humidity and low-temperature conditions (Radha et al., 1961). *C. gloeosporioides* causing infection develop more in high rainfall, relative humidity, and low-temperature conditions, and *Fusarium* spp. causing infection develop more in dry seasons (January-March) (Srinivasan and Gunasekaran, 1996b).

Management:

1. Removal of rotten portions of the spindle leaf and the adjacent two innermost fully opened leaves.
2. Dispensing of hexaconazole 5 EC @ 2 mL or mancozeb 3 g in 300 m of water around the base of the spindle leaf.
3. Apply the talc-based formulation of *Pseudomonas fluorescens*/ *Bacillus subtilis* @ 50 g in 500 mL of water by dispensing around the base of the spindle leaf. Treatments are imposed in all the root (wilt)-affected palms twice a year, that is, in April-May and

October-November. Treatment should not be carried out during rainy days (Srinivasan and Bharathi, 2006).

4. For avoiding red palm weevil damage, apply imidacloprid @1 mL/L of water by pouring the base of the spindle leaf.

4.4 PHYTOPHTHORA BUD ROT

Introduction/Economic Importance:

The disease was first observed in the Grand Cayman Island in 1834 (Tucker, 1926). Later, it was reported in India, Sri Lanka, West Indies, Fiji, Central America, Vanuatu, etc. In India, the disease was first reported by Radha and Joseph (1974) in Kerala at the incidence ranged from 0.1% to 6.5% and in Tamil Nadu at the incidence ranged from 0.4% to 6.7%. In Andhra Pradesh, the disease was observed in few gardens (ranged between 0.9% and 10%) at sporadic nature.

Symptoms:

The disease occurred in all the ages of palms, but young palms of 6-month seedlings to 10 years old are highly susceptible to the disease. The initial appearance of the disease symptom is seen in the spindle leaf which shows discoloration, followed by withering, and then turns into pale and becomes brown. On dissecting the affected trees, rotting of internal tissues could be observed and emitting a foul smell. The affected spindle leaves come off easily on slight pulling at this stage. It takes a few weeks for the decay of entire bud before destroying the meristem tissue. Water-soaked sunken lesions are observed in tender leaves, and the leaf that sheaths in early stages of the disease later turns to brownish (Radha and Joseph, 1974). In some cases, the infection leads to complete leave shedding, wilting, and death of the tree within the few months (Nambiar and Rawther, 1993). Brown to black necrotic patches with a yellow halo on the surface of the infected and mottled nuts appear in the internal portions. Younger nuts are highly susceptible; the affected nuts are failed to mature and drop off from the palms (Srinivasulu et al., 2008).

Casual Organism: *Phytophthora palmivora.*

The disease caused by the fungus *Phytophthora palmivora* that produces intercellular, nonseptate, hyaline mycelium, producing sporangiophores which are hyaline, simple, or branched occasionally. The sporangia are hyaline, thin walled, pear shaped with a prominent papilla which releases reniform, biflagellate zoospores. Further, it produces thick-walled, spherical oospores. In addition, it produces chlamydospores that are thick walled and yellowish brown.

Disease Cycle:

The pathogen has different types of spores such as sporangiospores, zoospores, chlamydospores, and oospores that may directly or indirectly cause infection. Once the life cycle has completed, it will take minimum of five weeks for the first appearance of the symptoms (Radha and Joseph, 1974). Sporangia are produced on infected nuts, leaves, stems, or roots, able to germinate directly to produce small, swimming zoospores. Sporangia and zoospores are spread by soil and soil water, wind-blown rain, and rain splash, whereas oospores and chlamydospores can survive in soil and coconut debris.

Epidemiology:

The disease becomes more severe in October, November, and December months because these are the months of more relative humidity, which is an important factor prevailing fungal infection and development of the disease (Menon and Pandalai, 1958). Some microclimatic factors such as high rainfall, atmospheric humidity (97%-100%), low temperature (18 °C-24 °C), injury, and the damage of Rhinoceros beetles (*Oryctus rhinoceros*) also influence the development of the disease (Radha and Joseph, 1974). During summer months, the fungus occurs in infected tissues as dormant nature and also survives at crop residues in the soil in the form of clamydospores and oospores (Menon and Pandalai, 1958).

Management:

1. On the disease diagnosis at the initial stages, curative measures can save the palm. The affected portions are cleaned and applied with 1% Bordeaux paste or Bordeaux mixture as premonsoon spray and the treated spindle portions covered with polythene sheet to prevent the entry of rainwater until new shoot emergence (Menon and Pandalai, 1958).
2. In later/advanced stages of infection, cut and remove the infected portions and burn them. In prophylactic measure, spray the plant with 1% Bordeaux mixture for all the surrounding healthy palms. Spraying mainly to be done with the spindle leaf and to the base of 2-3 innermost leaves.

3. Pour the copper oxychloride (0.25%) solution at the spindle leaf portion (Surulirajan et al., 2014).

4.5 GANODERMA BUTT ROT/BASAL STEM ROT/BOLE ROT/ TANJORE WILT/GANODERMA WILT

Introduction/Economic Importance:

It was first recorded on the coconut palm in Karnataka by Butler (1913) and first noticed in the severe form in the Thanjavur district, Tamil Nadu, India, after the cyclones of 1952 and 1955; hence, it was named as the Thanjavur wilt. The basal stem rot (BSR) disease is caused by *Ganoderma lucidum* (*G. lucidum*) (Leys) Karst. is a major limiting factor and most destructive disease in the coconut plantations, also known as bole rot, *Ganoderma* disease, Thanjavur wilt, and Anabe roga in different states of India. Now, it is widespread in Tamil Nadu, occurring in all districts. Infection spreads to neighboring palms by root contact if one palm becomes infected. The incubation period of the disease lasts for several years, and the disease symptoms appear only at a very late stage of infection. The management practices were found effective only when the disease was detected in the early stages (Bhaskaran et al., 1996).

Symptoms:

The palms in the age of 10-30 years old are easily attacked, and the fungus is soil-borne in nature and infects the roots. The *Ganoderma* pathogen first infects the root system, leading to extensive rotting, discoloration, decay, death of the fine roots, and becomes watery with a distinct smell of alcohol (Srinivasulu et al., 2002) which precedes the appearance of bleeding patches, that is, the exudation of reddish-brown viscous fluid from the basal portions of the stem. It gradually spreads upward even up to the height of several meters, rotting internally, leading to discoloration. In the advanced stage of infection, stem basal portion decays completely, resulting in breaking off from the root system. The sporophore appears in the base portion of the stem or from the roots just prior to the death of the palm (Bhaskaran et al., 1989). The leaflets droop and dry in outer whorls in the crown region, and subsequently, the other leaves also droop down in quick succession, leaving the unopened spindle leaf alone. They remain hanging around the trunk for several months before detaching. The spindle also breaks and falls off in the final stage of the disease, leaving the decapitated stem. The nut yield gradually declines, the production of new spathe stops, and the development of nuts in the existing bunches is also affected, leading to shedding of immature nuts (Bhaskaran, 2000).

Casual Organism: *Ganoderma lucidum* or *Ganoderma zonatum.*

Ganoderma wilt was first described under the name of *Fomes lucidus* (*F. lucidus*) (Leys) Fr. (Butler, 1909). The synonym for *G. lucidum* is *F. lucidus* wilt. The mycelium is aerial, hyaline, thin walled, branched with frequent clamp connections, and 1.4-2.9 μm in diameter; chlamydospores are abundant, slightly thick walled, ellipsoid, both terminal and intercalary, and 8.8-11.8 × 3.7-5.9 μm in size; cuticular cells from the crustose layer are hyaline to light brown, round to irregular in shape, and closely packed (Govindu et al., 1983).

The fungus produces a macroscopic basidiocarp which remains attached to the palm. The sporophores/brackets are very big with about 10-12 cm diameter and hardy. The upper surface is tough, shiny, and light to dark brown with concentric furrows, and the corresponding lower surface is white and soft with numerous minute pores which represent the opening of the hymenial tubes, which are lined with basidia and basidiospores.

Basidiospores are brown, thick walled, minutely verrucose, and truncate at one end, measuring 8.3-10 × 5.8-6.7 μm in size (Adaskaveg and Gilbertson, 1988). Basidiocarp production was observed in the sawdust medium two months after inoculation. The fungus was observed to grow in a wider range of pH. The maximum growth of the fungus occurred at pH 5.5 in culture. Glucose was the best carbon source for the fungus and peptone served as the best nitrogen source for *G. lucidum* (Nambiar and Radhakrishnan Nair, 1973).

Disease Cycle:

The fungus is soil borne and survives long time in soil, and in the summer, millions of spores are released into the air. These get blown into new places and are able to germinate as primary hypha if the conducive weather factor prevails. This new primary hypha must find a mate to continue on in the growth cycle with the process plasmogamy, meaning that the cytoplasms fuse together, that is, the dikaryon state (n + n). The fungus cannot produce the secondary hypha to help with growth, and the secondary hypha produces a very strong decomposing chemical called hydrolase to help break down the wood material. The secondary hypha can continue to grow for many years as long as the wood source/material is adequate. If the conditions become unsuitable for optimal growth, the fungus has to produce a fruiting body to release sexual spores.

Epidemiology:

The *Ganoderma* disease is severe in sandy or sandy loam soils in coastal areas of Tamil Nadu. The disease spread was lesser during the rainy season and completely inhibited by water stagnation (Srinivasulu et al., 2003). Hard pan formation in the subsoil impedes root penetration, which in turn predisposes the palms for infection (Ramasami et al., 1977). The *Ganoderma* wilt incidence was more between March and August (summer months) when soil temperature was the highest. The disease incidence was directly related to the mean maximum soil temperature, and it was not related to the minimum temperature, rainfall, and relative humidity (Bhaskaran et al., 1985).

Management:

1. Higher doses of NPK fertilizers can be avoided.
2. Soil application of calcium sulfate + magnesium sulfate each at 500 g/palm/year was found to be effective.
3. Intercropping with banana was found to be the most effective.
4. The frequent irrigation during summer reduces the intensity of the disease.
5. Soil application of quality neem cake @ 5 kg/palm reduces the disease intensity.
6. Root feeding with the fungicide hexaconazole @ 2 mL/100 mL water along with the soil application of neem cake was found effective.
7. Soil drenching with 1% Bordeaux mixture @ 40 L near the palm at the quarterly interval for thrice a year, which can be repeated after 2-3 years.
8. Application of aureofunginsol 2 g + copper sulfate 1 g in 100 mL of water through stem injection or root feeding at quarterly intervals for one year.
9. The soil application of bioconsortia (BS 1 mixture) having endophytic *Bacillus* (EPC 5) + *Pseudomonas fluorescens* strain (Pf1) + *Trichoderma viride* (Tv1) @ 100 g each along with 5 kg FYM/palm @ the three-month interval effectively reduced the disease severity with a higher nut yield.

4.6 STEM BLEEDING

Introduction/Economic Importance:

The disease was occurred in all coconut-growing areas in the tropical regions and was first reported from Sri Lanka (Petch, 1960) and India

(Sundararaman, 1925). In the beginning stages of the disease, there was not a large amount of yield loss. But in late stages, there was a stable yield decline, causing substantial defeat and in highly developed stages the death of the affected palms (Nambiar and Sastry, 1988b) occurs. The disease was more in sandy and laterite soils, especially in seashore or backwater areas.

Symptoms:

Stem bleeding is recognized by reddish/dark-brown spots, and sap flow down the stem. The color changes over time, becoming darker and turning black. These reddish/brown streaks correspond to a yellow to reddish color of the internal tissues of the stem. Stem cavities appear. The rot may extend to a large part of the stem, much in depth and height. The leaves droop and hang down the stem, and the nuts fall. Natural remission can occur, but the stem may break.

As a result of these symptoms, cavities are formed from the trunk from which liquid comes out when the bark is pressed or punctured. In the crown, the outer whorl of leaves become yellow, rather prematurely, dries up, and finally droop around the crown region. In later stages, fall of nuts is noticed when the palms are exposed to drought conditions. The trunk portion is slowly tapered in the direction of the apex and the crown size is reduced. Crown symptoms are more prominent during the summer and not in winter months in well-maintained gardens. In severe infection, the palm vigor is affected, and the nut yield is reduced. In extreme cases, trees may become unproductive and die (Nambiar and Sastry, 1988a; Warwick and Passos, 2009).

Casual Organism: *Ceratocistis paradoxa* (Dade) C. Moreau (synonym = *Ceratocystis paradoxa* Dade); the anamorph is *T. paradoxa* (de Seyn.) Höhn.

It is caused by *T. paradoxa*; the fungus is spherical and dark green and produces two types of conidia, namely macroconidia and microconidia. Macroconidia are produced on conidiophores singly; in chains, they are produced endogenously. Microconidia (endoconidia) are thin walled, hyaline, and cylindrical. The fungus also produces hyaline perithecia with a long neck, the base is ornamented with knobbed appendages, and the ostiole is covered by numerous pale-brown, erect, tapering hyphae. Asci are clavate, and ascospores are hyaline and ellipsoid.

T. paradoxa is isolated from affected tissues of the palms and is first recorded in the perithecial stage. This fungus produces pale-brown to brown-color hyphae. Conidiophores are slender, arising laterally from the hyphae, and produce cylindrical to oval-shaped endoconidia. When mature, they are changed to hyaline of pale brown with a smooth wall. The main characters of the chlamydospores are as follows: they form chains terminally, are of obovate to oval shape, thick walled, and brown. The perithecial stage is *Ceratostomella* (also known as *Ceratocystis*) *paradoxa* (Dade) Moreau; Perithecia are partly immersed, light brown, and produce numerous appendages of long, black neck and tapering, osteolar, hyaline; ascospores are ellipsoid, often with unequally curved sides, hyaline, nonseptate, and smooth, measuring 7-10 × 2.5-4 μm (Nag Raj and Kendrick, 1975).

Disease Cycle:

The fungus survives in infected plant debris, soil as perithecia, and conidia, and it spreads through wind-borne conidia. Further, the irrigation and rain-water also help in the disease spread. The beetles which feed on the diseased plants also help in the disease transmission.

Epidemiology:

The fungus is a very weak in nature and enters the trunk through wounds/cracks. Shallow loamy soils or laterite soil with clay or rock layer beneath the soil, poor drainage, poor maintenance of gardens, and damages done by *Diocalandra* and *Xyleborus* beetles, copious irrigation or rainfall, followed by drought are the favorable factors for the disease development (Sulladmath and Shantappa, 1980). The fungus survives in the infected plant debris and in soil in the form of perithecia and conidia. Irrigation and rainwater also help in the disease spread mainly through wind-borne conidia and beetles which feed on the diseased palms.

Management:

1. Severely infected palms can be removed and destroyed.
2. Bordeaux paste or burgundy paste should be applied in the infected portion from where the bark should be removed.
3. Basin irrigation can be done for each palm separately.
4. Soil application of 50 kg FYM, 5 Kg neem cake/palm/year.
5. Soil application of 50 g of *Pseudomonas fluorescens* and 50 g of *Trichoderma viride* along with 10 kg of FYM/palm at once in six-month intervals (Meena et al., 2014).
6. Root feeding with the fungicide hexaconazole 2 mL/100 mL of water thrice at three-month intervals to manage the disease.

4.7 ROOT (WILT) DISEASE

Introduction/Economic Importance:

Root (wilt) disease (RWD) is an imported disease in coconut plantations, and it affects various plantation crops, namely areca nut, date palm, and oil palm. The disease was first reported in the Erattupetta village of Kottayam district in Kerala state, India, and it is fast spreading in all the districts of Kerala and near the border districts of Tamil Nadu, namely Coimbatore, Kanyakumari, Theni, and Tirunelveli. The disease was first observed in the Cumbum Valley, Tamil Nadu, in 1971 and reported by Srinivasan et al. (2001). Since 1882, the coconut root (wilt) has been referred to as the "root disease" probably due to the rotting of roots (Butler, 1908). Nagaraj and Menon (1955) felt that name "wilt" would be a more appropriate term for the malady, taking into account the nature of foliar symptoms, and subsequently, the disease came to known as RWD.

Though it is not a lethal disease, it causes reduction in the quantity and quality of the nuts. If the palms are affected by this disease at the seedling stage, flowering is delayed, necrotized, and the yield is also considerably reduced. In affected plantations, the extent of the yield decline is proportional to the intensity of the disease which generally varies from 10% to 80% (Radha and Lal, 1972). Out of this loss, 43% occur in the early stage of the disease and 74% occur in the advanced stage. The incidence of the disease was 40.90% in bearing palms as compared to 16.80% in nonbearing palms. In India, it causes an estimated annual yield loss of 968 million nuts.

Symptoms:

Leaves:

Wilting and drooping of leaves, flaccidity, ribbing, paling/yellowing, and necrosis of leaflets are typical foliar symptoms (Radha and Lal, 1972). In general, the affected palms have 67%-97% of flaccidity, 38%-67% of yellowing, and 28%-48% of marginal necrosis. In younger palms (<10 years), it has 98.6% of flaccidity symptoms, while yellowing and marginal necrosis are nearly absent. Foliar yellowing and marginal necrosis of the older leaves were observed in association with disease, and in advanced stages, yellowing of younger leaves occurred (Radha and Lal, 1972) Holmes (1965) pointed out that such affected leaflets were curved along the entire length and formed a structure resembling the ribs of mammals. Dwivedi et al. (1978) observed that initial symptoms were manifestations of softening and whitening of the leaflets of the spindle, tip of leaves break, hang, dry, and then fall off. The sudden appearance of bright yellowing of 3-4 leaves in the middle whorl followed by the appearance of large number of brown spots of various shapes with a halo around all leaflets is the first symptom in certain cases. Unopened pale-yellow leaflets of spindle leaves are more susceptible to leaf rot and are a major component in the root (wilt), contributing to rapid decline and reduction in the yield. The secondary infection of the root (wilt) symptom is the leaf rot disease, which is caused by *Exerohilum rostratum* and *Colletotrichum gloeosporioides*, which occurs as superimposed on root (wilt)-affected palms. The leaf rot disease causes reduction in the photosynthetic area, disfiguration of the palms, and reduction in the yield, apart from attracting a number of insects that feed, multiply, and cause further damages (Chandra Mohan and Peter, 2008).

Roots:

Rotting of roots is considered to be one of the major symptoms (Butler, 1908). The percentage of the root decay varied from the intensity of disease at 12%-94.4% (Nagaraj and Menon, 1955). The anatomy of the roots of diseased palms revealed the degenerated phloem, disorganized tracheal elements, and tyloses in metaxylem (Govindankutty and Vellaichamy, 1983).

Inflorescence:

Flowering is delayed when palms are affected by the disease during the prebearing period, and the vitality of the reproductive system is very much affected (Radha and Lal, 1972). The spadixes become small, weak, and do not open normally. And drying of spath and necrosis of spikelets from the tip to downward occurs. Shedding of immature nuts and poor quality of nuts/copra often attribute to yet another character of the disease (Menon and Pandalai, 1958).

A number of physiological abnormalities such as water relations, mineral nutrition, respiration, photosynthesis, and phenol metabolism were noticed in the root (wilt)-affected palms. Rajagopal et al. (1986) observed that impaired stomatal regulation resulted in excessive water loss and in the occurrence of structural changes in vascular tissues and tyloses in xylem vessels of roots in diseased palms. In infected coconut palms, severe disturbances in the phloem transport and general physiological/biochemical changes occur (hormonal imbalances, respiratory rate, and vascular sap constituents). The uptake and transport of water through the trunk in diseased palms was reduced to 35% less than that of healthy palms (Ramadasan, 1964).

Causal Organism:

The disease is caused by phytoplasma and belongs to mollicutes which is cell wall-less prokaryotes and they are bounded by a "unit" membrane. The disease was transmitted by the plant hopper (*Protista moesta*) and the lace wing bug (*Stephanitis typica*) (Rajan and Mathen, 1985).

Disease Cycle:

Insect vectors are able to transmit the phytoplasma after salivary glands become infected at the 2-6-week latent period between acquisition and transmission. Phytoplasma was not observed in lace bugs collected from the disease-free areas such as Kasaragod in Kerala and Minicoy in Lakshadweep. Phytoplasma was observed in *Proutista moesta* salivary glands with an acquisition plus incubation period of more than 30 days on diseased palms. The inability of vectors to acquire from a particular plant species may be due to plant metabolites which disrupt insect feeding (Bosco et al., 1997). Spread is faster among the palms grown in sandy loams or alluvial soils. Higher disease incidence has been observed in low-lying water-logged areas bordering rivers and canals.

Management:

1. Disease-advanced juvenile palms, uneconomic palms yielding less than 10 nuts/palm/year can be removed.
2. Application of antibiotic, oxytetracycline, and temporary remission of the disease symptoms was observed (Srinivasan, 1999).
3. Growing of disease-resistant/tolerant varieties, namely Kalphasree (selection from Malayan Green Dwarf) and Kalpha Sankara (selection from the Chowghat green dwarf) (Chandra Mohanan and Peter, 2008).
4. Soil application of the balanced dose of chemical fertilizers (urea: 1.3 kg; superphosphate: 2 kg; muriate of potash: 3.5 kg/palm/year), 1 kg magnesium sulphate + organic manure: FYM @ 50 kg + *Pseudomonas fluorescens* 100 g + neem cake 2 kg/palm/year (Chandra Mohanan and Peter, 2008).
5. Mulching the basin with coconut leaves.
6. Growing green manure crops, namely cowpea, sunhemp (*Crotalaria juncea*), *Calopogonium mucanoides*, and *Pueraria phaseoloides*

may be sown in coconut basins during April-May and incorporated during September-October months.

7. Irrigation can be done during the summer month (250 L/day), and simultaneously proper drainage can also be provided.
8. Growing suitable inter, mixed crops (banana, pepper, cocoa, vanilla, turmeric, ginger, pineapple, coffee, nutmeg, or tapioca).
9. To manage the vectors (green lace wing bug and plant hopper), apply 20 g phorate and mix it with 10 g/200 g fine sand around the base of the spindle leaf or spray the dimethoate 1.5 mL + 1 mL sticking agent dissolved in 1 L of water for two times at the monthly interval.
10. To manage the leaf rot, remove the rotten portions of the spindle leaf and adjacent area and pour the fungicide hexaconazole 5 EC 2 mL or mancozeb 3 g in 300 mL of water around the spindle leaf.

4.8 CADANG-CADANG AND TINANGAJA DISEASES

Introduction/Economic Importance:

Cadang-cadang is derived from the Bicol term "gadan-gadan" which means dead or dying. It refers to the premature decline and death of coconut palms in Philippines associated with viroid infection. Similarly, tinangaja was first reported from Guam in 1917. By 1946, it had destroyed the coconut industry in the island (Boccardo, 1985). It is also associated with a viroid (*Coconut tinangaja viroid*, CtiVd) which is 64% homologous. *Coconut cadang-cadang viroid* (CCCVd) is the casual organism of the cadang-cadang disease (Boccardo, 1985; Keese et al., 1988). It is not reported in India till now.

Symptoms:

There are three main stages in the development of disease, early (E), mid (M), and late (L) (Randles et al., 1997). The E0 stage is detected in the youngest leaves, but plants are symptomless, whereas after 1–2 years, E1 stage develops as newly formed nuts become more rounded and exhibits equatorial scarifications without leaf symptoms. At the E2 stage, more nuts are rounded and scarified, chlorotic spots appear on leaves, inflorescences are stunted with tip necrosis, and loss of some male florets is observed. At the E3 stage, leaf spots enlarge, fewer nuts are produced, and new inflorescences are stunted and sterile. Further, spathe, inflorescence, and nut production decline, and then they cease by the M stage. In addition, leaf spots become more numerous, and by L stage, leaves decline in size and number, and leaflets become brittle. Also, leaf spots coalesce, resulting in general chlorosis, reduced crown size, and ultimately the death of the palm occurs. In 22-year-old palms, this averages 7.5 years, whereas in 44-year-old palms, the average is 15.9 years (Zelazny and Niven, 1980). Symptoms of tinangaja differ slightly from those of cadang-cadang (Hodgson and Randles, 1999).

Casual Organism: *Coconut cadang-cadang viroid* (CCCVd).

Pathogen:

Cadang-cadang and tinangaja are caused by viroids—small, naked, single-stranded circular RNAs. CCCVd was first isolated by Randles (1975). It has a basic 246 nucleotide structure that consists of a central conserved region of 44 nucleotides, which is common to many viroids (Haseloff et al., 1982). It predominates in the early stages of infection but is replaced by larger forms (287–301 nucleotides) as the disease progresses to the mid- and late stages. The *coconut tinangaja viroid* (CtiVd) has a sequence of 254 nucleotides (Keese et al., 1988), and the differences in the sequences of CCCVd and CTiVd could account for the variation in the symptom expression.

Transmission:

The mode of spread in the field is not known, and no insect vector has been found. Positive transmission was observed through the assisted pollination of mother palms with pollen from diseased palms.

Management:

1. Removing all the palms that were positively indexed for CCCVd infection might be more successful suggested by Randles et al. (1997).
2. Strict enforcement of quarantine regulations by concerned government agencies on the safe movement of coconut germplasm from infected areas.

4.9 COCONUT LETHAL YELLOWING DISEASES

Introduction/Economic Importance:

Recurrent lethal yellowing (CLYD)-type diseases infect coconut cultivation in the Eastern Hemisphere and include the diseases such as Awka disease (Nigeria), Cape St Paul wilt (Ghana), Kaïncopé disease (Togo), Kribi disease

(Cameroon), lethal decline (Tanzania, Kenya, and Mozambique), Kalimantan wilt, Natuna wilt and Sulawesi yellows in Indonesia, leaf scorch decline in Sri Lanka, Malaysian wilt in peninsular Malaysia, and Tatipaka disease and root wilt in India (Eden-Green, 1997).

Symptoms:

The typical coconut CLYD-type disease symptoms start with premature nut dropping, blackening (necrosis) of new inflorescences, followed by progressive yellow discoloration from the most basal to the youngest leaves. Further, rotting of the stem apical tissue (heart), wilting, crown collapsing within 3–6 months of initial symptom appearance (Hunt et al., 1973). These symptoms can be confused with other palm diseases, mainly those caused by the fungal BSR disease (*Ganoderma zonatum*) or by the abiotic deficiencies of boron (early nut fall), or potassium (discoloration and the early death of older leaves) (Broschat et al., 2010).

Casual Organism: Phytoplasmas.

Phytoplasmas are pleomorphic, cell-wall-less bacterial parasites with a bead-like, filamentous, or multibranched appearance under the electronic microscope (Seemuller et al., 2002; Oshima et al., 2013). Phytoplasmas share a two-host life cycle, involving plants and insect vectors (Garnier et al., 2001; Christensen et al., 2005). In the Caribbean, LY-type phytoplasmas have been detected in grass (Brown et al., 2008). The causal agent of coconut LY-type diseases has also been observed in Bermuda grass (*Cynodon dactylon*) and oil palm (*Elaeis guineensis*) (Nejat et al., 2009).

Transmission:

The most common phytoplasma transmission mechanism is dependent on phloem-feeding insect vectors which harbor and spread to different plants in a plant–insect–phytoplasma relationship (Garnier et al., 2001). Vectors are mostly in leafhoppers (Membracidae and Cicadellidae), planthoppers (Delphacidae, Derbidae, Cixiidae, and Flatidae), and Psyllidae (Weintraub & Beanland, 2006; Philippe et al., 2007).

Management:

1. The primary approach is often prevention that includes the use of resistant varieties, controlling the insect vectors, alternative plant hosts, and clearing out and destroying infected plants (Garnier et al., 2001).

KEYWORDS

- **coconut disease**
- **etiology**
- **symptomatology**
- **management**

REFERENCES

Abad, R. G. and Blancaver, R. C. (1975). Coconut leaf spot/blight and their control PCA-ARD. *Annual Report, 1975-1976.*

Adaskaveg, J. E. and Gilbertson, R. L. (1988). Cultural studies of four North American species in the Ganoderma lucidum complex with comparisons to *G. lucidum* and *G. tsugae*. Mycol. Res. **92**: 182–191.

Anonymous, (1979). Nematodes. fungi insects and mites associated with the coconut palm. Tech Bull No.2. Central Plantation Crops Research Institute, Kasaragod 670 124, Kerala, India, p. 236.

Anonymous (1985). *Coconut Root (Wilt) Disease: Intensity, Production Loss and Future Strategy*. Kasaragod: Central Plantation Crops Research Institute, p. 45.

Bhaskaran, R. (2000). Management of basal stem rot disease of coconut caused by *Ganoderma lucidum.* In: Flood, J., Bridge, P. D. and Holderness, M. (eds) *Ganoderma Diseases of Perennial Crops*. Wallingford, UK: CAB International, pp. 121–128.

Bhaskaran, R, Rajamannar M, and Kumar S. N. S. (1996). Basal Stem Rot Disease of Coconut, Technical Bulletin No 30. Kasaragod, Kerala, India, Central Plantation Crops Research Institute, 1996, p. 15.

Bhaskaran, R, Rethinam, P. and Nambiar, K. K. N. (1989). Thanjavur wilt of coconut. J. Plant Crops **17**(2):69–79.

Bhaskaran, R. and Shanmugam, N. (1985). Integrated pest and disease management. Proceedings of National Seminar. Coimbatore: Tamil Nadu Agricultural University. Problems and priorities in the management of Thanjavur wilt of coconut; pp. 183–187.

Boccardo, G. (1985). Viroid aetiology of Tinangaja and its relationship with cadang-cadang disease of coconut. In: Maramorosch, K. and McKelvey, J. J. (eds) *Subviral Pathogens of Plants and Animals: Viroids and Prions.* Orlando, FL: Academic Press, pp. 75–99.

Bosco, D., Minucci, C., Boccardo, G. and Conti, M. (1997). Differential acquisition of chrysanthemum yellows phytoplasma by three leafhopper species. *Appl Expt Entomol.*, **83:** 219–224.

Broschat, T.K., Elliott, M.L. and Maguire, I. (2010). Symptoms of diseases and disorders. In: *A Resource for Pests and Diseases of Cultivated Palms*. Gainesville, FL: University of Florida, Identification Technology Program, USDA. Available from: http://itp.lucidcentral.org/id/palms/symptoms. [Accessed on March 18, 2016].

Brown, J. F. (1973). Disease of coconut in the British Solomon Islands. Plant Dis. Rep. **57:** 856–860.

Brown, S. E., Been, B. and McLaughlin, W. A. (2008). First report of the presence of the lethal yellowing group (16SrIV) of phytoplasma in the weeds *Emilia fosbergii* and *Synedrella nodiflora* in Jamica. *Plant Pathol.*, **57:** 56–58.

Butler, E. J. (1908). Report on coconut palm disease in Travancore. *Agric Res Inst Pusa Bull.*, **9:** 23.

Butler, E. J. (1909). Formes lucidus (Leys) Fr., a suspected parasite. Indian Forester **35:** 514–518.

Chandra Mohan, R. and Peter, K. (2008). Root (wilt) of coconut economically viable: An easy to practice integrated management. *Indian Coconut J.*, **8:** 12–14.

Christensen, N. M., Axelsen, K. B., Nicolaisen, M. and Schulz, A. (2005). Phytoplasmas and their interactions with hosts. *Trends Plant Sci.*, **10(11):** 526–535.

Dwivedi, R. S., Mathew, C., Michael, K. l., Ray, P. K. and Amma, B. S. K. (1978). Carbonic anhydrase, carbon assimilation canopy structure in relation to nut yield of coconut. In: *Indian National Science Academy Symposium on Photosynthesis and Productivity, Lucknow*, pp. 44–46.

Eden-Green, S. J. (1997). History, distribution and research on coconut lethal yellowing-like diseases of palms. In: *International Workshop on Lethal Yellowing-like Disease, Chatham, UK*, pp. 9–25.

Elliott, M. L., Broscha,t T. K., Uchida, J. Y. and Simone, G. W. (eds) (2004). *Diseases and Disorders of Ornamental Palms*. St. Paul, MN: American Phytopathological Society.

Espinoza, J. G., Briceno, E. X., Keith, L. M. and Latorre, B. A. (2008). Canker and twig dieback of blueberry caused by *Pestalotiopsis* spp. and a *Truncatella* sp. in Chile. *Plant Dis.* **92:** 1407–1414.

Garnier, M., Foissac, X., Gaurivaud, P., Laigret, F., Renaudin, J., Saillard, C. and Bové, J. M. (2001). Mycoplasmas, plants, insect vectors: A matrimonial triangle. *Comptes Rend Acad Sci.-III*, **324(10):** 923–928.

Govindankutty, M. P. and Vellaichami, K. (1983). Histopathology of coconut palms affected with root (wilt) disease. In: Nayar, N. M. (ed) *Coconut Research and Development.* New Delhi: Wiley Eastern Ltd., pp. 421–425.

Govindu H. C., Rao A. N. S. and Kesava Murthy, K. V. (1983). Biology of Ganoderma lucidum (Leys.) Karst. and control of Anabe Roga of coconut. (In) Coconut Research and Development, pp 325–332. Nyar N M (Ed). Wiley Eastern Limited.

Haseloff, J., Mohamed, N. A. and Symons, R. H. (1982). Viroid RNAs of cadang-cadang disease of coconuts. *Nature*, **299:** 316–321.

Henry Louis, I. (2002). *Coconut—The Wonder Palm*, pp. 206–18. Hi-Tech Corporation Raman-puthoor, Nagercoil.

Hodgson, R. A. J. and Randles, J. W. (1999). Detection of coconut cadang-cadang viroid-like sequences. In: Oropeza, C., Verdeil, J. L., Ashburner, G. R., Cardena, R. and Santamaría, J. M. (eds) *Current Advances in Coconut Biotechnology*. Dordrecht, The Netherlands: Kluwer Academic Publishers, pp. 227–246.

Holmes, F. O. (1965). Investigations on the etiology of coconut root (wilt) disease. *Report to the Government of India* (Report No.1958 of the UN, Rome), p. 13.

Hopkins, K. E. and McQuilken, M. P. (1997). Pestalotiopsis on nursery stock, in *HDC Project News* No. 39. East Malling: Horticultural Development Council.

Hunt, P., Dabek, A. J. and Schuiling, M. (1973). Remission of symptoms following tetracycline treatment of Lethal Yellowing-infected coconut palms. *Phytopathology*, **64:** 307–312.

HYPERLINK "https://www.researchgate.net/profile/Prasanth_Lakshmanan" Lakshmanan, Prasanth and HYPERLINK "https://www.researchgate.net/scientific-contributions/R-Jagadeesan-2113913084" Jagadeesan, R. (2004). Malformation and cracking of nuts in coconut palms (Cocos nucifera) due to the interaction of the eriophyid mite Aceria guerreronis and Botryodiplodia theobromae in Tamil Nadu, India Journal of Plant Diseases and Protection. 11(2): 206-207.

Johnson, I., Meena, B. and Rajamanickam, K. (2014). Biological management of leaf blight disease of coconut using rhizosphere microbes. *J Plantation Crops*, **42(3):** 364–369.

Keese, P., Osorio-Keese, M. E. and Symons, R. H. (1988). Coconut Tinangaja viroid: Sequence homology with coconut cadang-cadang viroid and other potato spindle tuber viroid related RNAs. *Virology*, **162:** 508–510.

Keith, L. M., Velasquez, M. E. and Zee, F. T. (2006). Identification and characterization of *Pestalotiopsis* spp. causing scab disease of guava, *Psidium guajava* in Hawaii. *Plant Dis.*, **90:** 16–23.

Lingaraju, S., Naik, S. T. and Sastry, M. N. (1987). Host parasite relationship of *Pestalotia palmarum* Cooke. *Plant Pathol Newsl.*, **5(1–2):** 10.

McQuilken, M. P. and Hopkins, K. E. (2004). Biology and integrated control of *Pestalotiopsis* on container-grown ericaceous crops. *Pest Manag Sci.*, **60:** 135–142.

Meena, B., Rajamanickam, K., Kumar, M. and Sathyamoorthi, K. (2008). Evaluation of fungicides and biocontrol agents against leaf blight disease of coconut caused by *Lasiodiplodia theobromae*. *J Plantation Crops*, 36(3):463–465.

Meena, B., Ramjegathesh, R. and Ramyabharathi, S. A. (2014). Evaluation of biocontrol agents and fungicides against stem bleeding disease of coconut. *J Plantation Crops*, **42(3):** 395–399.

Menon, K. P. V. and Nair, U. K. (1948). The leaf rot disease of coconut in Travancore and Cochin. *Indian Cocon. J.*, **1(2):** 33–39.

Menon, K. P. V. and Nair, U. K. (1951). Scheme for the investigation of the root and leaf disease of the coconut palms in South India. Consolidated final report of the work done from 8th March 1937 to 31st March 1948. *Indian Coconut J.*, **5**:5–19.

Menon, K. P. V. and Pandalai, M. M. (1958). *The Coconut Palm: A Monograph.* Ernakulam: Indian Central Coconut Committee, p. 384.

Nagaraj, A. N. and Menon, K. P. V. (1955). Observation on root decay in coconuts, its cause and its relation to the foliar symptoms of disease in the disease belt of Travancore-Cochin. *Indian Coconut J.*, **8(97):** 105–121.

Nag Raj, T. R. and Kendrick, B. (1975). *A Monograph of Chalara and Allied Genera.* Waterloo, Canada: Wilfrid Laurier University Press.

Nambiar, K. K. N. and Rawther, T. S. S. (1993). Fungal diseases of coconut in the world. In: Nair, M. K., Khan, H. H., Gopalsundaram, P. and Rao E. V. V. B. (eds) *Advances in Coconut Research and Development.* New Delhi: Oxford and IBH Publishing Co. Ltd, pp. 545–561.

Nambiar K. K. N. and Sastry K. R. C. (1988a). Stem bleeding disease of coconut: reproduction of symptoms by inoculation with *Thielaviopsis paradoxa*. *J Plantation Crops*, **14:** 130–133.

Nambiar, K. K. N. and Sastry, K. R. C. (1988b). Stem bleeding disease of coconut. Current status and approaches for control. *Philippine J. Coconut Stud.*, **13:** 30–32.

Nambiar, K. K. N. (1994). Diseases and disorders of coconut. In: *Advances in Horticulture, Vol. X—Plantation and Spice Crops Part 1,* pp. 857–82. Chadha, K L and Rethinam P (Eds). Malhotra Publishing House, New Delhi.

Nejat, N., Sijam, K., Abdullah, S. N. A., Vadamalai, G. and Dickinson, M. (2009). Phytoplasmas associated with disease of coconut in Malaysia: Phylogenetic groups and host plant species. *Plant Pathol.*, **58:** 1152–1160.

Oshima, K., Maejima, K. and Namba, S. (2013). Genomic and evolutionary aspects of phytoplasmas. *Frontiers Microbiol.*, **4:** 1–8.

Pandey, R. R. (1990). Mycoflora associated with floral parts of guava (*Psidium guajava* L.). *Acta Bot Sin.*, **18:** 59–63.

Petch, T. (1960). Diseases of the coconut palm. *Tropical Agric.*, **27:** 489–491.

Philippe, R., Nkansah, J. P., Fabre, S., Quaicoe, R., Pilet, F. and Dollet, M. (2007). Search for the vector of Cape Saint Paul wilt (coconut lethal yellowing) in Ghana. *Bull Insectol.*, **60(2):** 179–180.

Polomer, M. K. and Bentonio, P. A. (1982). Control of grey leaf spot disease of coconut with fungicide and potassium chloride. *Philipp J Crop Sci.*, **7(3):** 166–168.

Radha, K. and Joseph, T. (1974). Investigations on the bud rot disease (*Phytophthora palmivora* Butl.) of coconut. *Final Report, Pl 480 Scheme*, p. 32.

Radha, K. and Lal, S. B. (1972). Diagnostic symptoms of root (wilt) disease of coconut. *Indian J Agric Sci.*, **42:** 410–413.

Radha, K., Sukumaran, C. K. and Prasannakumari, T. O. (1961). Studies on the leaf rot disease of coconut. Fungal infection in relation to environmental conditions. *Indian Coconut J.*, **15:** 1–11.

Rahman, S., Adhikary, S. K., Sultana, S., Yesmin, S. and Jahan, N. (2013). *In vitro* evaluation of some selected fungicides against *Pestalotia palmarum* (Cooke.) causal agent of grey leaf spot of Coconut. *J Plant Pathol Microb.*, **4(9):** 1–3.

Rajagopal, V., Patil, K. D. and Amma, B. S. K. (1986). Abnormal stomatal opening in coconut palms affected with root (wilt) disease. *J Exp Bot.*, **37:** 1398–1405.

Rajan, P. and Mathen, K. (1985). *Proutista moesta* (Westwood) and other additions to insect fauna on coconut palm. *J Plantation Crops.*, **13:** 135–136.

Ramadasan, A. (1964). Physiology of wilt disease in coco-nut palms. In *Proc. 2nd Session on FAD Wkg. Ply. Cocon. Prod. Prot. and Process., Colombo*, pp. 257–272.

Ramasami, R., Bhaskaran, R. and Jaganathan, T. (1977). Epidemiology of Thanjavur wilt disease of coconut in Tamil Nadu. *Food Farming Agric.*, **9(6):** 147–148.

Randles, J. W. (1975). Association of two ribonucleic acids species with cadang-cadang disease of coconut palm. *Phytopathology*, **65:** 163–167.

Randles, J. W., Hanold, D., Pacumbaba, F. P. and Rodriguez, M. J. B. (1997). Cadang-cadang disease of coconut palm: An overview. In: Hanold, D. and Randles, J. W. (eds) *Report on ACIAR-funded Research on Viroids and Viruses of Coconut Palm and Other Tropical monocotyledons 1985–1993. ACIAR Monograph* 45 (electronically published).

Seemuller, E., Garnier, M. and Schneider, B. (2002). Mycoplasmas of plant and insects. In: S. Razin and Hermann, R. (eds) *Molecular Biology and Pathology of Phytoplasmas*. London: Academic/Plenum/Kluwer, pp. 91–115.

Srinivasan, N. (1991). Occurrence of coconut leaf rot in relation to root (wilt) disease. *Indian Cocon J.*, **21(10):** 14–18.

Srinivasan, N. (1999). Know about phytoplasma as pathogen in relation to coconut diseases and with reference to root wilt. *Indian Cocon. J.*, **30(7):** 3–6.

Srinivasan, N. and Bharathi, R. (2006). Biocontrol agents against pathogens of coconut leaf rot disease. *J Plantation Crops*, 36(3):435–438.

Srinivasan, N. and Gunasekaran, M. (1993). Fungi associated with leaf rot disease of coconut. *Indian Cocon J.*, **23(10):** 2–7.

Srinivasan, N. and Gunasekaran, M. (1996a). Field control of leaf rot disease of coconut with fungicides. *Coconut Res Dev J.*, **12:** 34–42.

Srinivasan, N. and Gunasekaran, M. (1996b). Incidence of fungal species associated with leaf rot disease of coconut palms in relation to weather and the stage of lesion development. *Ann Appl Biol.*, 12(3):433–449.

Srinivasan, N., Koshy, P. K., Amma, P. G. K., Sasikala, M., Gunasekara, M. and Solomon, J. J. (2001). Appraisal of coconut (wilt) and heavy incidence of this disease in cumbum valley of Tamil Nadu. *Indian Cocon J.*, **31(1):** 1–5.

Srinivasulu, B., Aruna, K., Krishna Prasadji, J., Rajamannar, M., Sabitha Doraisamy Rao, D. V. R. and Hameed Khan. H. (2002). Prevalence of basal stem rot disease of coconut in coastal agro ecosystem of AP. *Indian Cocon J. XXXIII* (Nov): 23–26.

Srinivasulu, B., Aruna, K., Sabitha Doraisamy and Rao, D. V. R. (2001). Occurrence and biocontrol of Ganoderma wilt disease of coconut in coastal agro ecosystem of AP. *J. Indian Soc. Coast. Agricul. Res.* **19**(1&2): 191–195.

Srinivasulu, B., Gautam, B., Sujatha, A., Kalpana, M., Vijaya Lakshmi, P., Pavani Rani, A., Satya Ratna Subhas Chandran, B. and Rama Krishna, Y. (2008). AICAP on Palms, HRS, Ambajipeta *Technical Bulletin, Bud Rot Disease of Coconut*, p. 18.

Srinivasulu, B., Rajamannar, M., Aruna K, Rao, D. V. R. and Khan, H. H. (2003). Sero detection of coconut basal stem root pathogens G. applanatum and G. lucidum. J. Plant. Crops **31**(2):62–64.

Sulladmath, V. V. and Shantappa, P. B. (1980). Aetiology of stem bleeding disease of coconut in semi-dry tract of Karnataka. *Current Res.*, **9:** 73–75.

Sundararaman, S. (1925). Preliminary note on coconut leaf rot of Cochin. *Yearbook, Madras Agricultural Department*, 1924, pp. 6–8.

Surulirajan, M., Rajappan, K., Satheesh Kumar, N., Annadurai, K., Jeevan Kumar, K. and Asokhan, M. (2014). Management of bud rot disease. *Int J Tro Agri.*, **32(3–4):** 415–418.

Tucker, C. M. (1926). Phytophthora bud rot of coconut palms in Puerto Rico. *J Agric Res.*, **32:** 471–498.

Uchida, J. Y. (2004). Pestalotiopsis diseases. In: Elliott, M. L., Broschat, T. K., Uchida, J. Y. and Simone G. W. (eds) *Diseases and Disorders of Ornamental Palms*. St. Paul, MN: American Phytopathological Society, pp. 27–28.

Warwick, D. R. N. and Passos, E. E. M. (2009). Outbreak of stem bleeding in coconuts caused by *Thielaviopsis paradoxa* in Sergipe, Brazil. *Trop Plant Pathol.*, **34(3):** 175–177.

Weintraub, P. G. and Beanland, L. (2006). Insect vectors of phytoplasmas. *Annu Rev Entomol.*, **51(1):** 91–111.

Xu, L., Kusakari, S., Hosomi, A., Toyoda, H. and Ouchi, A. (1999). Postharvest disease of grape caused by Pestalotiopsis species. Ann Phytopathol Soc Jpn **65**:305–311.

Zelazny, B. and Niven, B. S. (1980). Duration of the stages of cadang-cadang disease of coconut palm. *Plant Disease*, **64:** 841–842.

CHAPTER 5

IMPORTANT DISEASES OF COFFEE (*COFFEE ARABICA* L.) AND THEIR MANAGEMENT

JAYALAKSHMI K.[1*], RAJU J.[2], RAGHU S.[3], and PRITI S. SONAVANE[4]

[1]*AINRP (Tobacco), ZAHRS, University of Agricultural and Horticultural sciences, Shimaoga, Karnataka 577201, India*

[2]*Plant Quarantine station, Ministry of Agriculture and Farmers Welfare, Government of India, Mangalore, Karnataka 575011, India*

[3]*Crop Protection Division, National Rice Research Institute, Cuttack, Odisha 753006, India*

[4]*Central Horticultural Experiment Station, ICAR-IIHR, Chettalli, Kodagu, Karnataka 571248, India*

[*]*Corresponding author. E-mail: jayalakshmipat@gmail.com.*

ABSTRACT

There are many species of coffee. Two important commercially grown species of coffee are: *Coffea arabica*, which produces Arabica coffee, and *Coffea canephora*, which produces Robusta coffee. Arabica covers about 60% of the world's coffee production, because of its higher quality, and can fetch much more in the market than Robusta. Warm and humid equatorial climates are better suited for Robusta coffee. It is generally more resistant to biotic and abiotic stress conditions than Arabica. Deep, free-draining, loamy soils, with a good water-holding capacity and a slightly acid soils (pH 5-6) are best suited for the growth of both the species. Soil fertility is important for good production. *Coffea* is a genus of flowering plants whose seeds, called coffee beans, are used to make various coffee beverages and products. Coffee plants can suffer from attacks by various diseases and they can affect quality, yield, and cost of coffee beans.

5.1 NURSERY DISEASES

Coffee is mainly propagated through seeds, and seedlings are raised in the nursery in two stages, namely germination beds and secondary beds (polybag). The nursery seedlings are mainly affected by three important fungal disease, namely collar rot, stem necrosis, and leaf spot and brown eye spot.

5.1.1 COLLAR ROT/DAMPING OFF

Introduction/Economic Importance:

It is the most devastating disease of coffee and can occur in an epidemic form under favorable weather conditions. Severe infection may lead to 10%-30% loss in yield in almost all growing areas (Green, 2011). The disease occurs in almost all parts of coffee-growing areas as well as in India, leading to severe reduction in yield (Reddy, 2004).

Symptoms:

The pathogen attacks both seeds and seedlings in two stages, namely pre-emergence and postemergence damping off:

1. Pre-emergence damping off: Fungus invades both embryo and endosperm of the seed before germination, and on account of this, the seed rots and disintegrates.
2. Postemergence damping-off: Discoloration at the collar region of the stem near the ground level (below the cotyledon) is seen in the seedling stage, leading to rotting of the tissue and death of seedlings.

Causal Organism: Collar rot or damping-off disease is caused by the fungus *Rhizoctonia solani* (*R. solani*) Kuhn.

Disease Cycle:

R. solani attacks seedlings of 1-4-month age, both in the germination beds and secondary polybag nursery. Under favorable hot humid weather conditions, loss of seedlings may go up to 10%-20%. Both Arabica and Robusta are susceptible to this pathogen. The incidence of collar rot is prevalent in all the coffee-growing regions. The fungus *R. soalni* is a soil inhabitant and attacks a wide host range. It can survive for a few years in the form of sclerotia in the soil and on plat debris. Sclerotia with thick outer layers survive and act as the overwintering spore for the pathogen. The pathogen also survives in the form of mycelium in the organic matter.

Epidemiology:

Excess soil moisture around the nursery beds on account of the poor drainage facility is favorable for the development of disease. Hot and humid weather conditions for a longer duration due to excess summer showers at frequent intervals are also favorable for disease development. Excess watering and overcrowding of seedlings favor the growth of the fungus and spread off the disease.

Management:

Following control measures should be practiced to reduce the severity of the disease incidence:

1. Site of germination beds in the nursery to be changed at least once in 2-3 years.
2. Sieved and sun-dried jungle soil, compost, and sand in the recommended proportion should be used for the preparation of seedbeds for filling of polybags with nursery mixture.
3. Treat seeds with carbendazim 50 WP at 1 g/kg before sowing.
4. Overcrowding of seedlings should be avoided.
5. Avoid excessive watering of the seedbeds.
6. Provide adequate filtered shade for the seedbeds and polybags (Adugna et al., 2007).
7. Remove and destroy the infected seedlings whenever observed. If infection is observed, spray 0.05% of carbendazim 50 WP or 0.4% of Dithane M-45.

5.1.2 STEM NECROSIS AND LEAF SPOT DISEASE

Introduction/Economic Importance:

Stem necrosis and leaf spot of coffee seedlings have been considered as one of the major diseases of coffee seedlings in Brazil, Columbia, Costa Rica, and Guatemala since the 1960s. There are reports of attack of *Myrothecium roridum* (*M. roridum*) on coffee seedlings from India. Earlier the problem was described in two different names, namely "target leaf spot" and "tip blight." Till 2005, this disease was rarely observed. However, since last

4-5 years, gradual increase in its incidence is seen, which is now causing 10%-15% mortality of seedlings in most of the coffee regions in Karnataka (Anonymous, 2003).

Symptoms:

In the seedling stage, the pathogen infects both stem and leaves. However, in some of the seedlings, the disease symptoms may be observed either on the stem alone, leaves alone, or on both leaves and stem.

On Leaves: The affected leaves show water-soaked circular spots at the initial stage, which gradually spread to more areas and change color from brown to black with concentric rings. Sometimes, two to three such lesions coalesce to form an irregular necrotic spot. In later stages, black fruiting bodies develop on under the surface of the affected leaves along the concentric rings.

On Stem: Water-soaked brown to black lesion can be noticed on any place of the stem above the collar region of the seedlings. On the infected lesion, the black fruiting body covered with white mycelia of the pathogen can be observed. Affected seedlings gradually start wilting and die.

At an early stage of disease development, symptoms appear similar to the "collar rot" disease. However, it can be differentiated from the collar rot disease due to the presence of necrotic lesions anywhere on the stem between the collar and tip of the seedlings.

Causal Organism: *Myrothecium roridum.*

Epidemiology:

Hot, humid weather, prolonged wetness of seedlings, low atmospheric temperature, and high relative humidity.

Management:

1. Avoid rising of germination beds in the same location for more than 2-3 seasons.
2. Use sun-dried sieved jungle soil, compost, and sand in the recommended proportion (6:2:1) while preparing either the germination of seedbeds or filling polybags for transplanting of seedlings.
3. Remove and destroy the affected coffee seedlings from the germination beds and in the polybags at regular intervals to limit the spread of the disease.

4. Protect seedlings by spraying propiconazole (tilt) 25 EC at 0.02% at monthly intervals from May to August (Hillocks et al., 1999).

5.1.3 BROWN EYE SPOT DISEASE

Introduction/Economic Importance:

Brown eye spot or leaf spot disease reported from all the coffee-growing countries including India. Coffee seedlings in the nursery and young plants in new clearings get infected when they lack proper overhead shade. The affected leaves drop, and such seedlings become unfit for field planting. Then both *Arabica* and *Robusta* seedlings are susceptible (Anonymous, 2009b).

Symptoms:

Affected leaves show circular necrotic spots with the dark-brown margin and light brown or pale center around the necrotic spot. The color of the leaf changes from green to yellow necrotic spots that increase in size, and the central portion turns light gray due to sporulation by the pathogen. During the rainy season, the central core portion of the spot disintegrates and collapses, leaving a shot hole at the middle of the spot. Infected leaves turn yellow and fall prematurely. Severely infected coffee seedlings show stunted growth and do not attain the desired growth.

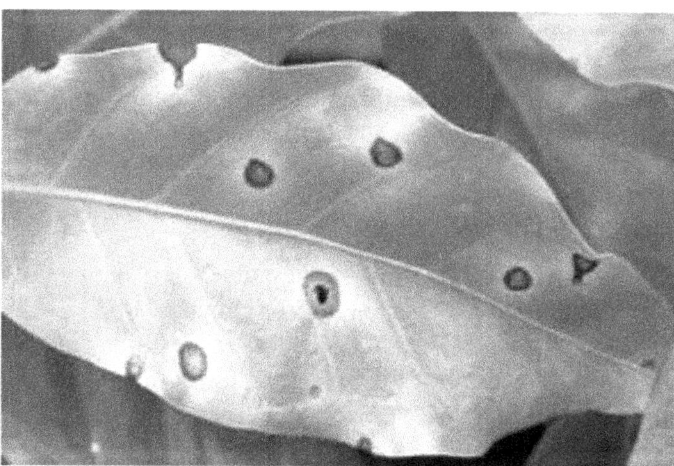

FIGURE 5.1 Typical symptoms of brown eye spot.

Causal Organism: *Cercospora coffeicola.*

Disease Cycle:

1. Primary source of inoculum: Dormant mycelium in infected plant debris, infected seeds, and volunteer plants.
2. Secondary source of inoculum: Air-borne spores.

The *Cercospora* spores (conidia) are borne on the surface of the leaflets and petioles and are disseminated by air currents, rain splash, and flowing water, on farm equipment and tools, and the workers' clothing. Hyphae from germinated spores penetrate through stomata and infect the foliage.

Epidemiology:

Sudden exposure of nursery seedlings to sunlight due to inadequate overhead shade and hot humid condition is the preconditioning factor for the development of disease. Favorable weather condition produces conidia, and dispersal takes place through wind and rain splash. Conidia germinate and penetrate through sun-injured tissue/stomata, leading to disease development. The pathogen survives on fallen debris for 7-8 months. The mean temperature of 22.5 °C-23.5 °C, relative humidity of 77%-85%, more sunshine hours (>5 hour/day), and a greater number of rainy days favor the disease. Disease development is favored by intermittent rains (Chapman, 2005).

Management:

1. Use only sun-dried sieved jungle soil, well-composed cattle manure, and sand in the recommended proportion of 6:2:1, while preparing either germination beds or filling polybags.
2. Grow coffee seedlings in the nursery by providing filtered light, preferably using 50% poly shade net.
3. Seedlings can be protected in the nursery by spraying one of the fungicides, namely Dithane M-45 or Dithane M-45 @ 0.4% (5 g/L) or carbendazim 50 WP @ 0.05% (1 g/L) at a 30-day interval.
4. During the first year of field planting, protect the plants from exposure to direct sunlight by providing hutting. Spray either carbendizim 50 WP @ 0.05% or Bordeaux mixture 1% if incidence is noticed. Do not use copper-based fungicides or Bordeaux mixture on coffee seedlings.

5.2 MAIN FIELD DISEASES

5.2.1 COFFEE LEAF RUST

Introduction/Economic Importance:

Coffee leaf rust is one of the most important diseases of coffee. Among the two commercially cultivated species, Arabica coffee is more susceptible to leaf rust than Robusta coffee. Severity and damage of rust are more in those areas where Arabica coffee is grown below 1000 m mean sea level. Rust affects the yield by loss of vigor of the plants. In severely affected areas, loss foliage up to 50%, and berries up to 70% can occur (Ritschel, 2005).

The earliest reports of the disease obtained from the 1860s. It was first reported by a British explorer from regions of Kenya around Lake Victoria in 1861 from where it is believed to have spread to Asia and the Americas. Rust was first reported in the major coffee-growing regions of Sri Lanka (then called Ceylon) in 1867, and the causal fungus was first completely described by the British mycologist Michael Joseph Berkeley and his collaborator Christopher Edmund Broome after an analysis of specimens of a "coffee leaf disease" collected by George H. K. Thwaites in Ceylon. Berkeley and Broome named the fungus *Hemileia vastatrix*, *Hemileia* referring to the half-smooth characteristic of the spores and *vastatrix* for the devastating nature of the disease (Schieber, 1972). The disease was recorded in India in 1870, Sumatra in 1876, Java in 1878, and the Philippines in 1889 (Avelino et al., 2001).

Symptoms:

1. Pale yellow lesions develop on the lower surface of the leaves.
2. Spots later turn yellow to orange color with powdery mass of uredeospores of the pathogen.
3. Size of the spots gradually increases and coalesces to cover the entire surface of the leaves.
4. Premature defoliation predispose the bushes to die back.
5. Severely infected bushes lose their framework and reduce the vegetative growth, number of bearing branches, berry size, and yield.
6. Coffee trees may suffer premature or complete leaf drop in severe cases.

Casual Organism: *Hemileia vastatrix* Berk & Br.

The leaf rust disease is caused by the fungus *Hemileia vastatrix* Berk & Br. It is an obligate parasite that affects only coffee. It is one of the most devastating and widespread diseases of coffee worldwide, has wiped out coffee in Sri Lanka (Ceylon) in the year 1868, and introduced to India in the year 1879 (Winston et al., 2009).

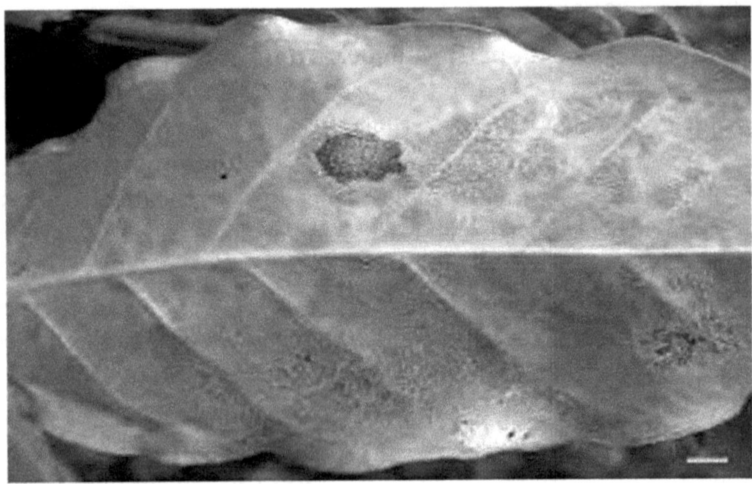

FIGURE 5.2 Coffee leaf rust.

Disease Cycle:

The pathogen forms uredeospores, teliospores, and basidiospores. But perpetuation is only through uredeospores formed abundantly on the infected leaves under favorable conditions. The pathogen exists in the form of races, and so far, 45 races of the pathogen are reported, of which 33 occur in India. Due to its ability to exist in the form of different races, newly developed resistant coffee varieties are subject to infection after a few years of commercial cultivation. Spread of the disease is by dispersal of spores, mainly by means of wind and water, and a little extent by other agencies such as insects, mites, and human beings.

Epidemiology:

1. Arabica coffee is mostly susceptible, while Robusta is resistant. Leaf rust can be especially devastating on trees under drought stress or where trees are insufficiently fertilized. Mulching and manuring of crop can help to minimize leaf rust impact. The disease-causing fungus itself may be attacked by other fungi. For example, *Verticillium* species parasitize the leaf rust fungus.

2. Intermittent rain and sunshine during June-September, ambient temperature from 22 °C to 24 °C, and relative humidity of 80% and above. Prolonged leaf wetness during the postmonsoon period either due to mist or intermittent Northeast rain. Thin overhead shade and overbearing disease development phases. Even though the incidence of leaf rust is observed on coffee bushes throughout the year, disease development is classified on the following four phases. Period of extension (April-August). After the blossom and backing showers, rust spots existing on the older leaves produce fresh spores. Disease build-up gradually increases on the older leaves as well as newly formed leaves during this period.
3. Period of intensification (September-November): Incidence reaches its peak during this period due to spread of the disease to the younger leaves formed on the bushes after the cessation of southwest monsoon (September).
4. Period of defoliation (December-January): Severely infected leaves with several rust spots generally shed leading to dieback of branches.
5. Period of inactivity (January-March): Due to low right temperature and humidity, spore production is almost arrested completely, and in addition, spores developed earlier also lose their viability due to unfavorable weather conditions. However, the pathogen survives on the infected leaves in the form of mycelia, and none of the fungicide have been found effective to kill the mycelia at this stage. Hence, fungicidal application is not required for control of disease during this phase.

Management:

1. Grow Arabica coffee under optimum shade consisting of a top canopy of recommended mixed shade trees and a middle canopy of dadap as temporary shade.
2. Prune the coffee bushed after harvest and before the blossom shower (January-February). Pruning during this period will help to reduce the inoculums of the pathogen.
3. Spray freshly prepare Bordeaux mixture (0.5%) with proper coverage of the under surface of the leaves before the set of the southwest monsoon (May–June).
4. Take up spot application triadimefon (Bayleton 25 WP @ 160 g/200 L) or hexaconazole (Contaf is a 5% EC at 400 ml/200 L) to highly susceptible plant populations of all the Arabica cultivars during break

in monsoon (August) to prevent defoliation and dieback (Hillocks et al., 1999).

5. Take up the second round of the prophylactic spray of the Bordeaux mixture @ 0.5% during the postmonsoon period (September–October). Susceptible cultivars such as 5.795 and Cauvery which show higher incidence during September should be sprayed with hexaconazole (Contaf @ 400 mL/200 L water) as the Bordeaux mixture spray will not contain disease and defoliation at this stage.

5.2.2 COFFEE BERRY DISEASE

Introduction/Economic Importance:

Coffee berry disease (CBD), which affects *Coffea arabica* (*C. arabica*), is caused by the fungus *Colletotrichum kahawae* (*C. kahawae*). It is endemic to Africa, and in 1922, it was first recorded in western Kenya. Since then, the disease has spread to most *C. arabica*-growing countries on the continent and is currently known to occur in Angola, Burundi, Cameroon, Central African Republic, Democratic Republic of Congo (DRC), Congo, Ethiopia, Kenya, Malawi, Mozambique, Rwanda, Tanzania, Uganda, Zambia, and Zimbabwe.

CBD can cause considerable yield losses of up to 75% when not adequately controlled (Reddy et al., 2003). Even though spray of fungicides to control the disease can result in yields being doubled, losses of up to 30% can still occur when attack is severe. Yield losses occur as a result of shedding and/or destruction of infected berries, which become mummified (dry, wrinkled and decayed, with a hard skin) (Carvalho, 1988).

Symptoms:

The characteristic symptom of CBD is the development of minute, water-soaked lesions on young, expanding berries that rapidly become dark brown or black and slightly sunken. The lesions enlarge to cover the whole berry within about a week, which subsequently rots.

Under humid conditions, masses of pale pink spores become visible on the surface of lesions. These symptoms, collectively known as "anthracnose," are typical of CBD. Another characteristic of the disease is the shedding of berries from branches at an early stage of disease development. Lesions may also occur on young berry stalks, causing them to be shed before lesions appear on the berry itself. The dark, sunken lesions are generally referred to as "active" lesions (Anonymous, 2011).

Casual Organism: *Colletotrichum kahawae.*

Kingdom:	Fungi
Phylum:	Ascomycotina
Class:	Sordariomycetes
Order:	Glomerellales
Family:	Glomerellaceae
Genus:	*Colletotrichum*
Species:	*C. kahawae*

Disease Cycle:

The fungus overwinters on the maturing bark of coffee twigs together with other species of *Colletotrichum*, such as *Colletotrichum gloeosporioides* and *Colletotrichum acutatum*. The maturing bark of coffee twigs has been shown to have the highest population of *C. kahawae* in comparison to immature, green bark, older bark, and berries. It constitutes, therefore, the primary source of inoculum for seasonal CBD outbreaks that develop after flowering and at the beginning of the rains, especially if no infected berries have remained on the tree.

Epidemiology:

Weather conditions are critical in the development of CBD. Adequate moisture is essential as the spores (conidia) of *C. kahawae* are dispersed by water and also require liquid water or 100% relative humidity for germination. This implies that CBD epidemics should be expected in areas where rainfall is generally high or during years of high rainfall in otherwise dry areas. Temperature is another important factor in that temperatures between 12 °C and 30 °C are also required for conidia to germinate, with the optimum being 22 °C. The host plant tissues may be infected within five hours of germination.

Management:

1. Avoid overcrowding of orchards. Wider spacing of plants will inhibit severe epidemics.
2. Providing wider spacing and ensuring that trees are pruned appropriately help to prevent prolonged wetness and high relative humidity, following periods of rainfall (conditions that are ideal for CBD development).

3. For effective management, there must be the adoption of proper sanitation practices. Tree sanitation is important. Infected twigs are to be pruned and burnt (along with fallen leaves).
4. Trees should be pruned yearly, and plant debris fallen on the ground need to be removed.
5. Trees may be sprayed twice with Bavistin (0.1%) at a 12-day interval during flowering to control blossom infection.
6. For the control of foliar infection, spraying of copper fungicides (0.3%) is recommended.

5.2.3 BLACK ROT DISEASE

Introduction/Economic Importance:

Black rot or Koleroga disease is considered as the second most important disease of coffee in India. Black rot is an endemic disease mostly found in areas which come under the influence of the southwest monsoon and receive annual rain fall above 80 inches. This disease is generally noticed in the valleys of the plantations. Black rot pathogen infects both the commercially cultivated species of coffee, but the crop damage caused by the pathogen varies from one place to another place and year to year depending on the extent and duration of rainfall. On severely affected plantations, the crop loss of 20%-30% can occur (Anonymous, 2012).

Symptoms:

1. Affected leaves, berries, and young shoots turn black and start rooting.
2. Disease development is more pronounced toward the center of the bushes.
3. Infected leaves get detached from the branches reveals the presence of mycelia running along the twig and leaf petioles and spreading on the lower surfaces of the leaves like a cub web.
4. Affected bushes, at the advanced stage, show defoliation, berry drop, and dieback, leading to the loss of entire framework.

Causal Organism: *Koleroga noxia* Donk.

Pathogen: *Koleroga noxia* Donk. (*K. noxia*) is a weak pathogen but under favorable conditions becomes virulent and causes severe damage within a short period of 8-10 days. *K. noxia* not only infects coffee but also several plants commonly observed in and around coffee plantations.

FIGURE 5.3 Black rot.

Disease Cycle:

This fungus has two distinct stages, namely pellicle and sclerotial in its life cycle. In the pellicle stage, the fungus produces white continuous mycelia amts on the affected leaves. Numerous basidiospores produced during this stage result in the secondary spread of the disease.

In the sclerotial stage, the host tissue turns brown to black with thick-walled and short-celled hyphal clumps of the pathogen. Such microsclerotia remain dormant on the host for 8-10 months and become active once the weather conditions are found favorable.

Epidemiology:

1. Continuous rain without a long dry spell.
2. Thick overhead shade.
3. Plantation in valleys and sheltered from sun light and wind.
4. Saturated atmosphere with 90%-100% relative humidity to continuous rain or hanging mist.
5. Overcrowding of coffee bushes due to closer planting.

Management:

1. Thinning of the overhead shade in the black rot endemic blocks before the onset of the southwest monsoon.

2. Minimize monoshade of silver oak in the diseas- prone areas, as the leaf fall covers the entire coffee canopy which is conducive for the incidence.
3. Remove criss-cross branches and tender shoots from the center, dead and dry branches, suckers, and fallen leaves of shade trees from the canopy of bushes before imposing premonsoon fungicide spray.
4. Ensure an adequate coverage of 1% Bordeaux mixture spray on both the surfaces of leaves and also on the developing berries just before the onset of monsoon.
5. Remove and destroy the affected leaves and berries along with the thread of mycelia to prevent further spread of the disease.
6. Spray carbendazim 0.03% (Bavistin 50 WP at 120 g/200 L) in black rot endemic areas after 45 days of the first spray of the Bordeaux mixture under the clear break of monsoon. Infected leaves and berries need to be removed from the bushes, before taking up of the fungicide spray.

5.2.4 COFFEE TRUNK CANKER

Introduction/Economic Importance:

Coffee canker disease caused by *Ceratocystis fimbriata* (*C. fimbriata*) was first reported from Indonesia (Java) in 1900 (Pontis, 1951). Further, this disease was reported from Colombia in 1932 from the Quindío Province. Currently, the disease is found in all coffee-growing areas of the world. The canker disease of coffee incited by *C. fimbriata* has been found for the first time in Costa Rica. The disease is characterized by chlorosis, defoliation, and eventual death of coffee stems, branches, and the tree itself.

Coffee canker disease is the most devastating disease which reduces the profitability of coffee crops in India. This disease caused more severity and more incidences in recent years, resulting in substantial losses to yield as well as economic loss to coffee farmers *(Winston et al., 2005).*

Symptoms:

1. Sunken bark with hard dry irregular lesion on the main stem.
2. Beneath the bark, brown to black discoloration of the internal tissue up to 5-7.5 mm depth.
3. Yellowing of leaves, wilting of branches, and death of the plant.

Causal Organism: *C. fimbriata.*

Disease Cycle:

The fungus survives in soil and on the decaying organic matter such as plant debris left in the field. It can survive for several years in the soil. Wounds on root are important entry points for infection by the fungus. Coffee roots and stems are susceptible to infection. The ascospores of the sporulating *C. fimbriata* have accumulated in a sticky drop at the tip of their perithecial necks. The combination of the fruity aroma and the sticky spores is thought to be an adaptation for promoting fungal dispersal by insect vectors. The fungus is mainly spread by wind, water, soil, farm implements, by some insects, by humans (clothing), by contaminated tools.

Epidemiology: Disease is more in heavy soil and increases with soil moisture.

Management:

1. Uproot and destroy severely infected plants.
2. Stump the plants which show the initial stage of infection.
3. Apply the Bordeaux paste to the cut end.

5.3 ROOT DISEASES

Introduction/Economic Importance:

Brown, red, black, and Santavery are the four different types of root diseases reported on coffee in India. Among the four, Red and Brown root diseases not only affect Arabica and Robusta coffee but also several common shade trees. Red and brown root diseases are reported from all the coffee-growing regions. The incidence of black root disease is rarely noticed in Arabica coffee. In the endemic areas, Santavery root disease causes considerable death of plants at its prime age of yielding.

Symptoms of Brown Root Disease:

1. Affected plants at the initial stage have lesser foliage. Leaves gradually turn yellow, branches start wilting followed by defoliation, leading to the death of the plant.
2. Stem near the ground becomes spongy and soft. Thick, brown encrustation of the pathogen is noticed on the tap root and secondary roots just below the collar region.

3. Close observation of the surface of the infected root reveals the presence of closely interwoven brown strands of mycelia of the pathogen. The inner portion of the root shows dark-brown to black tawny lines of fungal hyphae.

Causal Organism: *Fomes noxius.*

Symptoms of Red Root Disease:

1. Diseased plants show gradual yellowing of foliage, wilting of branches, and defoliation flowed by the death of the bush.
2. The root system of dying or dead coffee plants shows the red encrustation of the fungal mycelia covered with soil and gravel.
3. The surface of the root of the affected plants appears deep red in color when washed under running water.

Causal Organism: *Poria hypolateritia.*

Disease Cycle of Brown and Red Root Disease:

Infected host plant debris left over in soil is the most common source of primary inoculum in newly established orchards or plantations. Infected seedlings from the nursery can also serve as the initial source of inoculum in the field. The fungus spreads primarily via mycelial contact with roots. The fungus colonizes the root system and moves to the collar and may form encrustation on the trunk of the host. These continue to develop and liberate spores until the end of the rainy season when a layer of the sterile tissue seals the pores.

Epidemiology of Brown and Red Root Disease:

1. The fallen logs and shade tree stumps in an estate serve as a substrate for multiple pathogens which are generally present in the soil.
2. Soil with low pH, less organic content, and temperatures ranging from 10 °C to 35 °C favor the development of the pathogens.
3. Collateral hosts of brown and red root disease pathogens: Both the pathogens have a wide host range, and they not only attack coffee but also several other commercial crops such as tea, cocoa, nutmeg, oil palm, rubber, and citrus, and also many common shade trees of coffee plantations. *Greviliea robusta*, *Syzygium jambalana*, *Pterocarpus marsuplum*, *Dalbergia tatifalia*, *Ficus tjaketa*, *Ficus glomerata*, *Acracarpus fraxinifollus*, *Artacorpus integrifolia*, and *Erythrina lithosperma* are some of the shade trees that are affected by the red and brown root diseases.

5.3.1 SANTAVERY ROOT DISEASE

Introduction/Economic Importance:

Santavery root disease affects only Arabica coffee and is confined to certain localities of the eastern zone of Karnataka and the Yercaud region of Tamil Nadu.

Symptoms:

1. Sudden wilting, yellowing of leaves followed by the defoliation and death of aerial parts of plants.
2. Roots in the transverse section show brown to pinkish discoloration of inner portion. Scraping the bark of the stem near the ground level also shows internal discoloration.

Causal Organism: *Fusarium oxysporum* f. sp. *Coffeae* (*F. oxysporum*).

This pathogen is a soil-inhabiting saprophytic fungus which changes to the parasitic phase under extreme weather conditions that are ideal for the development of the pathogen but adverse for the growth of coffee plants (Adugna, 2007).

Disease Cycle:

F. oxysporum produces asexual spores (microconidia and macroconidia). It produces chlamydospores and overwinters as mycelium or spores in infected or dead tissues. It can spread by air, equipment, and water.

Epidemiology:

1. Poor physical condition of the soil with more pebbles and stones around the root system.
2. Change in soil temperature, either too low or high, for a long period and moisture stress.
3. Exposure of plants to high-speed wind causing wounds/eruption on the bark due to wind swaying.
4. Inadequate overhead shade.

Management:

1. Like other soil-borne pathogens, it is very difficult to eliminate the population of *Poria hypolateritia* and *Fomes noxicus* completely from the infected soil. Hence, importance should be given to avoid disease incidence.

2. Remove the fallen logs and shade-tree stumps to prevent the establishment and multiplication of the pathogens.
3. Maintain adequate overhead shade.
4. Adequate care should be given to keep the vigor of the coffee bushes by applying organic manure.
5. Protect the coffee plants from exposure to wind by developing a proper wind belt.
6. Remove and destroy the affected area and one row of apparently healthy adjoining plants all around with 60-cm-deep and 15-cm-wide trenches.
7. Apply 2-3 kg of agricultural lime and turn the soil in the pits from where affected coffee plants and shade trees are uprooted. Also, apply agricultural lime to the affected areas inside the trench and dig the cover during the dry period to expose the soil sunlight.
8. Exposing the pits to sunlight for 6-8 months before planting.
9. Drench the soil with carbendazim 50 WP (8 g/L) at 2-3 L/plant at the collar region twice in a year (pre- and postmonsoon) for treating root-disease-affected plants that show initial symptoms of infection. Also, the application of *Trichoderma harzianum* (125 g) with well-decomposed farm yard manure (FYM) (3 kg) twice in a year can reduce the disease.

5.4 NEMATODE PROBLEMS IN COFFEE

Introduction/Economic Importance:

Coffee being the most important crop of all the four states of South India suffers from many nematode diseases. Several pathogens of *Meloidogyne* spp. (*Meloidogyne exigua*, *Meloidogyne incognita*, *Meloidogyne coffeicola*, *Meloidogyne arenaria*, and *Meloidogyne jvanica*) have been reported from coffee plants. However, the pest of utmost significance is *Pratylenchus coffeae*. Of the two commonly grown species, *C. arabica* and *C. robusta*, the former is highly susceptible to nematodes, while the latter is either tolerant or resistant to nematode attack (Winston, 2009).

5.4.1 COFFEE ROOT LESION NEMATODE:

It is one of the highly destructive pests of economic significance in *C. arabica* in all coffee-growing areas.

Symptoms:

1. Infected plants bear lean and thin stems and give unhealthy appearance. Affected plants do not bear sufficient foliage to yield the crop.
2. Older leaves become yellow and drop off, leaving very few undersized, crinkled, and chlorotic leaves at the tip of the main stem.
3. Branches have shorter internodes, sparse flower buds, and poor setting, and beans produced are small in size and unfilled.
4. Even bearing plants have a lanky stem. Plants become prematurely old and require collar pruning (rejuvenation) or uprooting. Even collar-pruned plants do not produce new suckers. Even if they produce some suckers, they are unhealthy with reduced, crinkled, and chlorotic leaves.
5. Affected plants thus have a poor anchorage and can be uprooted easily.

Causal Organism*:* *Pratylenchus coffeae* infects roots of coffee plants causing damage to the cortical tissue.

Disease Cycle:

Lesion nematodes penetrate plant roots completely and migrate throughout the root tissue, mainly the cortex, as they feed. They can penetrate anywhere along the roots, but they show some preference for the region near the start of the root hair zone. They penetrate the root epidermis either intra- or intercellularly, but once inside, they migrate intracellularly. The nematodes feed on cells within the root, usually until the cells lyse and cavities are formed, and then the nematodes move forward within the root to feed on healthy plant cells. There are no elaborate plant cell modifications induced by lesion nematodes for feeding as there are many sedentary plant-parasitic nematodes. The migration of the nematode within the root is usually ahead of the developing zone of necrosis that culminate in a visible lesion.

Epidemiology:

In soil, lesion nematodes do not usually migrate more than 1-2 m from the root zone that they infect. In plantings where root grafts may occur (i.e., fruit trees), the nematodes may travel from plant to plant through roots. Areas of disease become more pronounced in adverse environmental conditions such as water and nutrient stress or if secondary pathogens simultaneously infect the roots. The spread of these nematodes within fields is usually accelerated by the cultural practices of the grower, such as soil cultivation. Lesion nematodes can be introduced to noninfected sites by poorly sanitized farm equipment and contaminated planting stock. Since many species of *Pratylenchus* are endemic to native vegetation in many locations, new planting sites may already be infested with the nematode prior to cultivation.

Management:

1. Dig soil deep to expose to the sun during summer.
2. Raise nematode-free seedlings in the nursery.
3. Destruct of weeds and change the nursery site once in every 6-7 years.
4. Avoid carrying soil from the infected area to healthy ones.
5. Uproot infected plants and burn.
6. Apply neem cake of 1 kg/plant two times in a year.
7. Rotate crops with soybean, cotton, maize, and castor.
8. Maintain heavy shade, apply FYM, foliar feed nutrients, and mulch individual plants help to improve the vigor of the plant.
9. Preplant soil treatment of the coffee nursery with carbofuran 3G, followed by plastic sheet covering or Dazomet application effectively controls *P. coffeae*.
10. Apply with carbofuran at 3 g/plant every three months at least for one year to protect the disease in the initial stages.
11. Graft coffee cultivars susceptible to the nematode (*C. arabica*) on resistant root stocks (*C. canephora* var. *robusta*) as this technique has also been successful in the control of *P. coffeae*.

KEYWORDS

- **coffee crop disease**
- **etiology**
- **symptomatology**
- **management**

REFERENCES

Adugna, G., Hindorf, H., Steiner, U., and Dehne, H. W. (2007). Mating test and in vitro production of perithecia by the coffee wilt pathogen, *Gibberella xylarioides* (*Fusarium xylarioides*). *Ethiop J Biol Sci.*, 6(1): 63–75.

Anonymous, (2003). *Central Coffee Research Institute (CCRI) (2003). Coffee Guide*. Chikmagalur, Karnataka, India: Directorate of Research, Central Coffee Research Institute, Coffee Research Station.

Anonymous, (2009b). Central Coffee Research Institute (CCRI). "Coffee Varieties from CCRI: Arabica." Extension Folder No.1/2009. Directorate of Research, Central Coffee Research Institute, Coffee Research Station, Chikmagalur, India. Government of Karnataka.

Anonymous, (2012). Government of India. (2012). *Annual Report 2010-2011*. Bangalore: Food, Civil Supplies and Consumer Affairs Department.

Anonymous, (2011). Kerala Agricultural University. (2011). *Package of Practices Recommendations: Crops*. 14th edition. Thrissur: Kerala Agricultural University, p. 360.

Avelino, J. H., Zelaya, A., Merlo, A., Pineda, M., Ordonez, and S. Savary. (2006). The intensity of a coffee leaf epidemic is dependent on production situations. *Ecol Model*., 197(3): 431–447.

Carvalho, A. (1988). Principles and practice of coffee plant breeding for productivity and quality factors: *Coffea arabica. Coffee: Agronomy*. Ed. R. J. Clarke. New York, NY: Elsevier Applied Science.

Chapman, K. (2005). *Arabica Coffee Manual for Myanmar*. Bangkok: FAO Regional Office for Asia and the Pacific.

Green, W. H. (2011). *Econometric Analysis*. New York, NY: Prentice Hall.

Hillocks, R. J., Phiri, N. A., and Overfield, D. (1999). Coffee pest and disease management options for smallholders in Malawi. *Crop Protect*., 18: 199-206.

Phiri, N. A., Hillocks, R. J., and Jeffries, A. (2001). Incidence and severity of coffee diseases in smallholder plantations in Northern Malawi. *Crop Protect*., 20: 325-332.

Pontis, R. E. (1951). A canker disease of the coffee tree in Colombia and Venezuela. *Phytopathology*, 41: 179–184.

Reddy, D. R. B., Shivprased, P., and Naidu, R. (2003). Estimation of cost of production of Arabica coffee in Chikmagalur region, Karnataka. *J Coffee Res*., 31(2): 106–118.

Reddy, D. R. B. (2004). Analysis of cost of production of Arabica coffee in Kodagu region, India. *J Plant Crops*, 32(3): 53–57.

Ritschel, A. (2005). Monograph of the genus Hemileia (Uredinales). *Bibliotheca Mycol*., 200: 1–132.

Schieber, E. (1972). Economic impact of coffee leaf in Latin America. *Annu Rev Phytopathol*., 10: 491–510.

Winston, E., de Laak, J. O., Marsh, T., Aung, O., Nyunt, T., and Chapman K. (2005). *Arabica Coffee Manual for Myanmar*. Bangkok: FAO Regional Office for Asia and the Pacific.

Winston, E., de Laak, J. O., Marsh, T., Aung, O., Nyunt, T., and CCRI. (2009). *Diseases of Coffee Leaf Rust* (Extension Folder No.17/2009). Chikmagalur: Directorate of Research, Central Coffee Research Institute, Coffee Research Station.

CHAPTER 6

IMPORTANT DISEASES OF TEA (*CAMELLIA SINENSIS* L.) AND THEIR INTEGRATED MANAGEMENT

PEZANGULIE CHAKRUNO, SUSANTA BANIK*, and KAVI SUMI

Department of Plant Pathology, SASRD, Nagaland University, Medziphema, Nagaland 797106, India

**Corresponding author. E-mail: susanta.iari@gmail.com.*

ABSTRACT

Tea is regarded as the most popular and inexpensive beverage produced from the shoots of commercially cultivated tea plants. It is grown in more than 50 countries with India as the largest producer of the world. During the period 2015–2016, India produced 1233 million kilograms of tea from an area of 567 thousand hectares in India, majority of the made tea is produced from northeast, particularly from Assam. Like any other plants, tea also suffers from several fungal foliar diseases. The perennial habit of the tea plant and other factors like cultural conditions and warm humid climate required for growing tea becomes highly favorable for the development of diseases. The following diseases are serious constraints to the production and productivity.

6.1 INTRODUCTION

Tea, *Camellia sinensis* (*C. sinensis*) (L.) O. Kuntze, is a perennial plantation crop which belongs to the family *Theaceae* (Bhujel et al., 2016). Tea is originally from the mountains of Southeast Asian countries. There are many varieties of tea but *C. sinensis* var. *sinensis* and *C. sinensis* var. *assamica* are the most popular (Lehmann-Danzinger, 2000). Regions receiving annual rainfall range from 1150 to 8000 mm and temperature range between -8 °C and 35 °C with a day length of 9.4-15 hours are the most favorable conditions for the cultivation of tea. And it can be cultivated in a wide range of soil that is acidic in nature (Natesan, 1989; Surendramohan, 1995). Tea is regarded as the most popular and inexpensive beverage produced from the shoots of commercially cultivated tea plants. It is grown in more than 50 countries with India as the largest producer of the world (Balamurugan et al., 2013). During the period 2015-2016, India produced 1233 million kilograms of tea from an area of 567 thousand hectares (Tea Board India, 2016). In India, majority of the made tea is produced from Northeast, particularly from Assam. Like any other plants, tea also suffers from several fungal foliar diseases (Almada-Rui et al., 2003) (Table 6.1). The perennial habit of the tea plant and other factors like cultural conditions and warm humid climate required for growing tea becomes highly favorable for the development of diseases (Baby, 2002).

TABLE 6.1 List of Tea Diseases

S. No.	Name of the Disease	Causal Organism
Major Diseases		
1.	Blister blight of tea	*Exobasidium vexans* Massee
2.	Red rust of tea	*Cephaleuros parasiticus*
3.	Brown blight or gray blight	*Pestalotia theae*
4.	Twig dieback, stem canker	*Macrophoma theicola*
Minor Diseases		
5.	Root diseases	Most of the root rotting fungi belong to the subdivision basidiomycotina and ascomyotina
6.	Dieback	*Nectria cinnabarina*
7.	Brown blight	*Coletotrichum camelliae*
8.	Horsehair blight	*Marasmius crinis-equi* *Marasmius equicrinis*
9.	Phloem necrosis	Phloem necrosisvirus (Camellia Virus 1)

Some of the important tea diseases causing severe economic losses are discussed below.

6.2 BLISTER BLIGHT OF TEA

Introduction/Economic Importance:

Blister blight disease in tea is a serious foliar disease as it directly deteriorates the quality of the made tea and severe crop losses in some cases. It is caused by an obligate fungus *Exobasidium vexans* (*E. vexans*) Massee (Baby et al., 1998). The disease is prevalent in Sri Lanka, Burma, Malaya, Indonesia, Japan, and Taiwan. It also appeared in 1912 in Japan and 1930 in Vietnam. In 1949, the blister blight jumped over to Sumatra, and from there, in 1950, to the neighboring island Java (Indonesia). Now, it is endemic throughout most tea-growing areas of Asia where it has reached epidemic proportions.

The disease was first time reported in India from Assam in 1868 at the border of Myanmar, where it caused little damage. Then it spread in 1908 to Darjeeling, and from there, in 1946, to south India and Sri Lanka (Venkataram, 1974). After that the disease spread in other tea-growing states of the country. Since then, spreading in India and Asia still remains an unresolved problem. In India, the disease occurs in most of tea gardens every year and causes severe damage to the crop (Peal, 1868).

The first epidemic on tea due to this disease was recorded in 1906 from upper Assam, India, that caused severe widespread destruction (Mann, 1906). It was believed that the disease was confined to Assam only until in 1908 it spread to Darjeeling, causing heavy losses to the tea industry (Me Rae, 1910). Since then, the disease has been reported from all tea-growing countries of Asia (Sowndhararajan, 2013a). Due to blister blight disease, heavy losses are incurred by the tea industry annually. Studies reported that unprotected (unsprayed with fungicide) tea fields indicated a loss of 50% in six months.

Symptoms:

The symptom of blister blight on tea first appears on the young leaves and later on the older leaves. Initially, the symptom appears as small, round, pale-green translucent spot on the leaf. In rare cases, deep-red spots may also appear; the disease also infects the young tender stem such that the leaves and buds above the diseased part become necrotic and wither away (Petch, 1923). The disease spread rapidly, and soon almost all plants are attacked, and within few days, after the infection, the spots enlarge and covert as blisters. The blisters may form on both the surfaces of the leaves. The spots on the leaf showing sunken stomates gives concave shape to the upper side and convex shape to the inner side. Later, as the disease advances, circular blisters start to appear on the inner side of the leaf with a dark-green water-soaked zone around, and then the blisters turn into white velvety on which spores are produced (Lehmann-Danzinger, 2000). The development of blister lesions on the mid rib and the margin may sometimes distort or fold the affected tea leaves (Holliday, 1980). Old spots become whitish or grayish in color and are covered by a whitish powdery coating.

Causal Organism: *E. vexans* Massee.

Blister blight of tea is caused by *E. vexans* Massee (Massee, 1898). It belongs to the division Basidiomycotina, class Hymenomycetes, in the order Exobasidiales of the subclass Holobasidiomycetidae (Lehmann-Danzinger, 2000).

The mycelium of fungus is confined to the blister areas on the leaves and tender stem region. The mycelium of fungus is septate inter- and intracellular. Before sporulation, the dikaryotic hyphae are collected in bundles between the epidermal cells of the host (Venkataram, 1970).

Disease Cycle:

The disease occurs mainly in the rainy season. Cool, moist, still air favors infections, as do higher elevations. According to Fuchs (1989), more than 4-5 hours of sunshine/day keeps the attack of fungus below the danger level.

Basidiospores of *E. vexans* after their landing on the tea leaf surface take about 24 hours to germinate; the entire life cycle of the pathogen is completed in 11 days under favorable weather conditions (de Weille, 1959a). Sugha (1997) reported that during the off (dry) season, the pathogen survives in the necrotic blister leaves, and upon the return of favorable conditions, the pathogen sporulates. It has been reported by Ajay et al. (2009) that *E. vexans* produces thick-walled basidiospores under adverse conditions for survival.

Epidemiology:

Abiotic factors such as higher relative humidity (>80%), low atmospheric temperature (<24 °C), longer duration of leaf surface wetness, with shorter sunshine duration greatly favor blister disease in tea (Ajay et al., 2009). It has also been reported that it attains epidemic proportion during the periods of the two monsoons in southern India (Ram, 1974). Survival of the fungus in the off season was on necrotic blisters. Fuchs (1989) reported that with the sunshine hours of more than 4-5 hours per day, the disease can be kept below the danger level. The disease gets more severe in those plants which are recovering from pruning since they are debilitated and have mainly young leaves (Lehmann-Danzinger, 2000).

Integrated Disease Management:

Cultural Method:

1. Removal of shade trees in the tea plantation field has been reported to reduce the severity of the disease (de Weille, 1959b; Liau, 1966).

Biological Method:

1. Sowndhararajan et al. (2013b) reported that *Ochrobactrum anthropi*, a phylloplane bacteria isolated from tea when applied as foliar spray, was found effective against blister blight disease of tea.

Chemical Method:

1. Thakur and Masand (2015) reported that Nativo (trifloxystrobin 25% + tebuconazole 50%) @ 125 g/hectare was effective against blister blight of tea.
2. Application of copper-based fungicide drastically reduces the incidence of blister blight disease in tea (Ram, 1973; Ram and Mouli, 1983).
3. Baby et al. (2004) reported that the continued application of ergosterol biosynthesis inhibitors provided some control over the disease by suppressing the spore germination, sporulation, and by reducing the spore size and viability.
4. Tea leaves are traditionally protected from the infection of *E. vexans* by the application of copper oxychloride (COC) at 125 g a.i./hectare or at a higher concentration after pruning. Since only the young leaves and tender twigs can be infected by the fungus, only these have to be sprayed with the fungicide. The layer of copper on the leaves will kill the germinating spore of *E. vexans*.
5. To control the blister blight in India, a mixture of copper fungicide and nickel chloride is sprayed at 7-10-day intervals throughout the rainy season.

6.3 RED RUST OF TEA

Introduction/Economic Importance:

The red rust disease is regarded as one of the most serious diseases in tea plants known to be caused by an alga *Cephaleuros parasiticus* (*C. parasiticus*). The incidence of the disease is highly prevalent in tea-growing countries such as India, Bangladesh, Sri Lanka, China, Japan, and Kenya. It was also reported that the severity of the disease varied from locality to locality (Baby, 2001). In India, the disease is most common in the tea plantations of southern India. This disease affects both the young and matured leaves (Mann, 1901; Mann and Hutchinson, 1904). In severe cases, the pathogen is fully capable of attacking the stem of tea plants. According to Rattan (1993), the disease is more serious in young tea fields, mostly attacking the mature leaves only.

Symptoms:

The symptoms of the disease are usually seen on the leaves of the affected tea plant (Chen and Chen, 1990). The disease appears as roughly circular, greenish gray, raised lesions on the affected leaves which turn to reddish brown. Reddish hairy spots are often found on the upper surface of the leaf, extending to the stem. Older infections become greenish gray and look like lichen (Keith et al., 2006). The severely affected leaves are discolored, and the shoots become pale so as to appear unhealthy in the case of stem infection, and in some cases, the stem tissues are killed causing severe damage to young tea plants (Huq et al., 2010).

Causal Organism:

Red rust disease of tea is caused by an alga *C. parasiticus* Karst (Sana, 1989; Huq et al., 2007) belonging to the family Trentepoholiaceae of the division Chlorophyta in the algal kingdom (Chapman, 1984).

The algal filaments on leaves are found in the epidermal cells and in between the palisade cell. On the stem region, the pathogens first attack the bark and then penetrate into the cortex. From the orange-red fruiting patches, which are formed under favorable conditions, fine, cottony hyphae grow consisting of sporangiophore and sporangia.

C. parasiticus is also able to cause infection in *Vanilla planifolia*. Both *C. parasiticus* and *Cephaleuros virescens* are referred to as parasitic on tea, mango, citrus, and litchi. *C. parasiticus* was once referred to as the parasitic state and *C. mycoidea* as the epiphytic or saprophytic state. However, the clear evidence of death of mesophyll cells up to the lower epidermis as a result of the failure of cork formation is taken as a strong indication of the pathogenic nature of the alga; thus, a separate identity of a saprophytic state is ruled out. To avoid this confusion and any serious, pending taxonomic enquiry, it might be prudent and pragmatic to refer to all the species tentatively as *Cephaleuros* spp., as reported by Joubert and Rijkenberg (1971).

In the initial stage of disease development, the microscopic examination of the affected portion of the leaf reveals the presence of a thick and compact mat of alga thallus beneath the cuticle. The thallus is disk-like and is composed of symmetrically arranged cells radiating dichotomously from the center to the periphery. At the center, the disk has several layers of cells, and at the periphery, it has only one. Usually, the parasite remains between the cuticle and the epidermis, and sometimes, it does extend up to the palisade layer. Very rarely, the very young thallus as on magnolia may be superficial. Intracellular penetration is, however, ill evidenced. Under the subcuticular infection in leaf lies a large air space between the alga disk and the necrotic epidermal cells of the host (Agnihothrudu, 1964).

The rhizoid or the cellular processes extend through the air space into the underlying host cells. This provides the necessary anchorage and feeding system for drawing water and nutrients. The ultrastructural studies show that the young cell walls of the alga, in direct contact with the host cuticle, are considerably thickened by the secondary wall material (Satynanarayana and Barua, 1983). The asexual reproductive structure of *C. parasiticus* is of sporangium type with a few sporangiophores, septate, and orange colored, mixed with setae that arise toward the periphery of the thallus disk after rupturing the cuticle or the epidermis.

The further growth of the vegetative cells shows sporangiophores appearing subterminally or even centrally from the disk. The sporangiophores measuring 105–3991 mm × 9.4–45 mm, mostly being 238.2 mm × 29.91 mm, terminate in a swollen apical cell, called vesicle. Usually, four or may be eight or even more protrusions or sterigmata come out from the vesicle, which are light orange, unicellular, and measure 16–45.5 mm × 5–15 mm. Each of them develops into an orange, oval to globose zoosporangium, measuring 19–30.4 mm × 19–22.8 mm, and is attached to the apex by a curiously folded stalk through the ostiole in a suitable temperature and moisture regime. The number of zoospores released per sporangium varies from 8 to 30 (Chapman, 1984; Ponmurugan et al., 2010).

The filaments of *C. parasiticus* are very thin, erect, and green in color, usually measuring 50.31–23 mm and 240.82 mm. The external thallus of this alga grows faster than those in the internal one. The external thallus bears numerous erect, sterile filaments, while fertile filaments bear clusters of 5–6 sporangia at the apex. Filaments' cells are usually reddish brown as they contain b -carotene. Sporangiophores are 7–8 septate, being distributed at an equal distance. The sporangia develop on about 15 days-old alga. They are oval to elliptical and measure 22.12 mm × 18.18 mm. The mature sporangia break away from the filaments and are dispersed with wind and rainwater. Biflagellate zoospores, produced under congenial humid conditions, germinate to form a new thallus under the favorable condition (Chapman 1984; Ponmurugan et al., 2010).

Sexual reproduction, as observed on tea, takes place in a homothallic manner, by the fusion of the isogametes, produced within large, sessile, flask-shaped cells. After fusion, the zygote produces a dwarf sporophyte consisting of an attaching stalk that bears one or more cells. Each cell has a small microsporangium. The meiosis completed, and each microsporangium releases four quadriflagellate microzoospores. In the presence of free water, the flask-shaped cells open by fissure and release the swarm spores (zoospores) reported by Chapman (1984) and Ponmurugan et al. (2010).

Disease Cycle:

The alga can survive in the stem and old leaves and become severe when bright sunlight is available and when the new flushes come out. *Cephaleuros parasiticus* produces small-microscopic reddish spore-like structures on the surface of the affected leaf. These spores are easily disseminated by rain splashes or by wind (Keith et al., 2006). Infected leaves and other plant parts serve as a sauce of infection. The survivability of the pathogen is well established by the availability of spores throughout the year, and most of them come from the large number of collateral hosts such as *Albizzae chinenisia*, *Teophosoria*, and *Candida deamodium* sp., are being present in abundance in the vicinity of tea plantations in the northeast India (Sharma, 1960b). The zoospores of *Cephaleuros parasiticus* which serve as the inoculum of the disease are produced abundantly during the rainy season prior to which the infection severely occurs right after the rainy seasons. Sporangia repeatedly formed on the primary growth act as a source of secondary infection (Chapman and Henk, 1985).

Epidemiology:

Red rust has the ability to attack both young and old tea plants under adverse conditions of soil and climate. The development of red rust disease in tea plants is highly influenced by factors such as fertility, alkalinity (low pH) and poor soil aeration, hard pan, absence of shade, drought, and water logging (Sharma, 1960a). The severity of the disease is highly governed by the climate and ecological conditions. According to Hasan et al. (2014), shading conditions of the tea plants showed significant impact on disease development. The incidence of the disease is lesser in shaded areas than that of unshaded ones. The pathogen can infect the branches at any stage of growth. Once the branch is infected, the algae remain latent for a year (Huq et al., 2007).

Integrated Disease Management:

Cultural Method:

1. Culture practice like deepening of drainage, low pH, and nutritional deficiencies should be identified and corrected (Joubert and Rijkenberg, 1971; Ponmurugan et al., 2006; Bhujel et al., 2016).
2. Promote good air circulation in the plant canopy to reduce humidity (Keith et al., 2006).

3. Huq et al. (2010) reported that the application of muriate of potash (K) as a potash fertilizer @ 210 kg/hectare greatly reduced the severity of red rust and increased the yield of made tea.

Chemical Method:

1. Bordeaux mixture at 3:3:100 or a copper fungicide with 50% metallic copper also gives good control against tea red rust under regular plucking of tea leaves (Baby, 2001).
2. Dutta and Barthakur (1998) recommends applications of 0.25% COC beginning from the end of April at fortnightly intervals for the first two sprays, followed by two sprays at monthly intervals.

6.4 BROWN BLIGHT OR GRAY BLIGHT OF TEA

Introduction/Economic Importance:

Gray blight is also considered as one of the important foliar diseases of tea (*C. sinensis*). It is becoming a serious problem in almost all parts of tea-growing regions in India, particularly in tea-growing districts of southern India. During peak seasons, that is, from July to December, the crop loss due to gray blight disease blight is estimated to be 17%, and both the quality and quantity of tea are affected (Premkumar et al., 2012; Kumhar et al., 2016).

Symptoms:

The disease mostly affects the mature leaves of tea, and then infection progresses to the young shoots and the tender foliage resulting in huge crop losses (Horikawa, 1986; Sanjay and Baby, 2005; Joshi et al., 2009). Initially, the disease appears as small pale yellow-green, oval spots on the leaves. Later, the spots turn from brown to gray in color, forming concentric rings with scattered; tiny black dots become visible and eventually necrosis, leading to defoliation (Keith et al., 2006).

Causal Organism:

Gray blight disease in tea is caused by *Pestalotiopsis theae* (Ram, 1983; Premkumar et al., 2012) which belongs to the division Ascomycota, class Sardariomycetes, order Xylariales, and family Sporocadaceae.

The mycelium of fungus is septate, much branched, and is mostly intercellular. Haustoria arising from the intercellular mycelium enter the cells of tissues and absorb nutrition required for the pathogen.

Disease Cycle:

Espinoza et al. (2008) made a report that either asexual conidia or fragmented spores of *Pestalotiopsis theae* are responsible for the infection. Fungal spores are transported from the infected tea plants or site of infection to the healthy ones by rainwater splash (Keith et al., 2006). The pathogen could well survive harsh weather conditions and may cause primary infection upon the return of favorable weather conditions (Premkumar et al., 2012).

Epidemiology:

Pestalotiopsis theae are considered as weak pathogens (Madar et al., 1991) which usually affect tea plants that are already weakened by adverse environmental conditions or improper aftercare. High temperature, humidity, and poor air circulation favor the infection (Keith et al., 2006). Abiotic stress greatly favors the extent of the disease (McQuilken and Hopkins, 2004).

Integrated Disease Management:

1. Grow tea bushes with adequate spacing to permit air to circulate and reduce humidity and the duration of leaf wetness (Keith et al., 2006).
2. Anita et al. (2011) and Kumhar et al. (2016) reported that the foliar spray of carbendazim 12% a.i. + mancozeb 63% a.i showed maximum control over the disease under *in vitro* conditions.
3. *Pseudomonas* consortium integrated with carbendazim followed by Nativo 75 WG (15 g in 10 l) and the integration of *Trichoderma* consortium with carbendazim provided satisfactory control over the disease (Premkumar et al., 2012).
4. Barman et al. (2015) reported an *in vitro* test among chemical fungicide. Bavistin showed 100% inhibition over *Pestalotiopsis theae*, while *Trichoderma viride* showed 74.3% inhibition of mycelial growth over the control.

6.5 TWIG DIEBACK, STEM CANKER

Introduction/Economic Importance:

This disease is capable of reducing yields and can kill entire plants. Rainy weather favors its spread, and dry conditions promote its development. The disease, which is most serious during the dry summer, caused the death of more than 40% of tea bushes in certain orchards.

Symptoms:

The first symptoms include browning and drooping of affected leaves. As the disease spreads into the shoots, they become dry and die. The entire branch can die from the tip downward. Dying branches often have cankers shallow, slowly spreading lesions surrounded by a thick area of bark (Keith et al., 2006).

Causal Organism: *Macrophoma theicola.*

Disease Cycle:

The fungus produces spores on small, pear-shaped pycnidia on dead branches. Spores are spread when splashed by rain and can survive for several weeks on pruned branches left in the field. The fungus usually requires wounded plant tissues to gain entry and initiate infection (Keith et al., 2006).

Integrated Disease Management:

1. Plant in well-draining, acidic soils.
2. Remove diseased twigs by cutting several inches below cankered areas and disinfecting them.
3. Spray appropriate protective fungicides during periods of wet weather or natural leaf drop to protect leaf scars from infection.

6.6 HORSEHAIR BLIGHT

Introduction/Economic Importance:

Disease is occurred worldwide including India.

Symptoms:

Black fungal threads resembling horsehair are attached to upper branches and twigs by small brown disks. The fungus penetrates and infects the twigs from the disks and produces volatile substances that cause rapid leaf drop.

Causal Organism: *Marasmius crinis-equi* (also *Marasmius equicrinis).*

Disease Cycle:

This pathogen is spread from infected twigs to healthy twigs by extending its hair-like threads.

Integrated Disease Management:

1. Remove and destroy all crop debris from around plants.
2. Prune out infected or dead branches from the plant canopy.

6.7 ROOT DISEASES

Introduction/Economic Importance:

Hainsworth (1952) reported that 1% loss of bushes of tea were involved with root diseases, but it would increase 1% every year. Hence, the potential crop expected in the 10th year would be 10%. Root diseases are the limiting factor of the tea production in many countries.

6.7.1 RED ROOT ROT

Symptoms:

Wilt of the plant and dieback. Roots with white surface mycelium later turn red, to which soil adheres. Rot appears pale brown and hard, and later becomes pale buff and either dry or spongy. The general symptoms of root diseases are a dieback of twigs and branches of the bushes, and a wilt of the seedlings.

Causal Organism: *Ganoderma pseudoferreum* (Wakef.) van Overh.et Steinm; *G.philippi* (Basidiomycotina, Hymenomycetes, Aphyllophorales, and Ganodermataceae).

6.7.2 BRICK-RED ROOT ROT

Symptoms:

Root surface white speckled with mycelial strands. These turn into a smooth sheet hardening into plates or ropes of red color. They show up with scraping or washing. The general symptoms of root diseases are a dieback of twigs and branches of the bushes, and a wilt of the seedlings.

Causal Organism: *Poria hypolateritia* Berk (Basidiomycotina, Aphyllophorales, and Polyporaceae). Also, *Ganoderma pseudoferreum* (Basidiomycotina, Hymenomycetes, Aphyllophorales, and Ganodermataceae).

6.7.3 BLACK ROOT ROT

Symptoms:

Rot of root and stem base, black wood discoloration. White mycelia on roots, later gray to black. Stem ringing. *R. bunodes* has a wide host range. The general symptoms of root diseases are a dieback of twigs and branches of the bushes and a wilt of the seedlings.

Causal Organism:

Rosellinia arcuata Petch; *R. bunodes* (Berk. et Broome) (Ascomycotina, Pyrenomycetes, Sphaeriales, and Xylariaceae).

6.7.4 CHARCOAL STUMP ROT OR CHARCOAL ROOT OR USTULINA CHARCOAL ROT

Symptoms:

No surface mycelium. Under bark of roots white fan-like patches. Wood at the base of the stem with irregular double lines. A wide range of host. The general symptoms of root diseases are a dieback of twigs and branches of the bushes, and a wilt of the seedlings.

Causal Organism: *Ustulina deusta* (Hoffm.) Lind; *U. zonata* (Lév.) Sacc. (Ascomycotina, Pyrenomycetes, and Sphaeriales).

6.7.5 DIPLODIA ROOT ROT OR DIPLODIA

Symptoms:

Dieback. Root and stem-base rot. The blackened vascular system and black discoloration of wood. A large number of host plants. The general symptoms of root diseases are a dieback of twigs and branches of the bushes, and a wilt of the seedlings.

Causal Organism: *Lasiodiplodia theobromae* (Pat.) Griffon et Maubl. (syn. *Botryodiplodia theobromae* Pat.). (Deuteromycotina. Coelomycetes). Pantropical Aarmillaria root rot or root-splitting disease.

Symptoms:

Sudden browning of leaves. Root splitting. White mycelial mat under the bark of the stem base and roots, shoe-string rhizomorphs. The general symptoms of root diseases are a dieback of twigs and branches of the bushes, and a wilt of the seedlings.

Causal Organism: *Armillariella mellea* (Vahl:Fr.) (syn. *Armillaria mellea*) (Basidiomycotina, Hymenomycetes, Agaricales). Severe in Africa and Indonesia, in India rare.

6.7.6 *WHITE ROOT ROT*

Symptoms:

Dieback, white mycelium (rhizomorphs) on root. Roots rotten and white. Living wood decaying fungi. The general symptoms of root diseases are a dieback of twigs and branches of the bushes, and a wilt of the seedlings.

Causal Organism: *Rigidoporus lignosus* (Klotzsch) Imazeki (Basidiomycotina, Hymenomycetes, Aphyllophorales, and Polyporaceae).

6.7.7 *BROWN ROOT DISEASE*

Symptoms:

Adherence of a crust of earth and gravel round the entire root. White or brown mycelium under the bark. Pale dry rot of wood. The general symptoms of root diseases are a dieback of twigs and branches of the bushes, and a wilt of the seedlings.

Causal Organism: *Phellinus noxius* (Corner) Cunn. (syn. *Fomes noxius* Murr.) (Basidiomycotina, Aphyllophorales, and Hymenochaetaceae). Africa, SE Asia, Australasia

6.8 RHIZOCTONIA SEEDLING BLIGHT

Symptoms:

Seedlings wilt and dieback. Wide host range. Soil-inhabiting fungi. The general symptoms of root diseases are a dieback of twigs and branches of the bushes, and a wilt of the seedlings.

Causal Organism: *Rhizoctonia bataticola* (Taub.) Butl. (Deuteromycotina, Agonomycetes); teleomorph probably *Tanatephorus cucumeris* (Basidiomycotina, Hymenomycetes).

Epidemiology:

Soils with low pH, low content of organic matter, and previous clearing from more or less degraded forests are the factor for disease development.

Disease Cycle:

In soil, other soil-inhabiting microorganisms keep the population of the facultative pathogens low. A higher diversity and population density of the saprophytic soil microorganisms will reduce the population of the facultative pathogens accordingly, thereby lowering the chance of the facultative pathogens to attack healthy tea bushes. However, facultative root pathogens have very favorable growth conditions on root and wood residues in the soil, remaining from former trees and bushes of the site. These residues give rise to locally high populations of the facultative pathogens, meaning a high inoculum density. On such sites, the facultative root pathogens can easily attack healthy roots of tea bushes.

Integrated Disease Management:

1. First, removal of stumps, other wood residues and large roots of woody plants from the soil prior to planting. This applies also to shadow trees removed from the plantation.
2. The second strategy is increasing the diversity and population density of the soil microflora, by augmenting the organic matter of the soil either with green manure, with appropriate shadow trees, or by adding organic manure such as compost.
3. Woody plant residues should be destroyed by burning or burying them beyond reach of roots in the soil.
4. Controlling the root and dieback diseases appearing in the plantation is done by first removing the diseased bush and all neighboring tea bushes together with their roots and destroying them by burning or burying them outside the plantation.
5. Replanting is possible only after a two-year fallow with nonhost plants, such as grass. If after two years there are still appreciable wood residues remaining in the soil, a one-year treatment of the replanted tea bushes with systemic fungicides is recommended.
6. Suitable systemic fungicides are the group of demethylation inhibitors (DMI4), such as bitertanol, flusilazole, propiconazole, penconazole, hexaconazole, and others, and in the group of morpholines tridemorph5. In Sri Lanka, good results have been observed after the application of this technique.

7. Obtained by drenching systemic fungicides in 3-4-month intervals for a period of one year (Arulpragasam et al., 1987). The application of fungicides of the same group is consecutively avoided to not promote the resistance of the fungus against the fungicide. In any case, the application of organic manure (compost) is recommended before replanting.

6.9 PHLOEM NECROSIS OF TEA

Introduction/Economic Importance:

Phloem necrosis is a virus disease that affects seedling tea, causing a general debilitation of the tea bush with an eventual decline in yields to uneconomic levels.

Symptoms:

As the name implies, this disease is caused by the death of the phloem tissues due to the effect of the virus. The earliest symptom of disease is the occurrence of necrosis that affects the phloem of roots. Later, necrosis develops in the aerial part of the plant. The external symptoms are the leaf curl associated with a zigzag habit of stem growth, shortening of the internodes, and dwarfing of the shoots. Severely affected bushes become entirely unproductive.

Roots: The necrosis is readily demonstrated by shaving away a slice of bark of the root so that the phloem is exposed close to the cambium. If the bark of the root is examined, the occurrence of dead zones of tissues, which are brownish in color, would be noticed. The phloem in the brownish tissue is dead; hence, there is the occurrence of discoloration. It is important to note that necrosis is not continuous but occurs in irregular patches. Microscopic examination reveals brownish zones of the dead phloem tissue.

Stems: Regions of the discolored dead tissue similar to that observed in the roots are found in the bark of the stems, though phloem necrosis in the stems is not as widespread as in the roots.

Leaves: Dwarfing of the leaves is the general symptom of the disease. Lower surfaces of the petiole and the midrib of the leaf are most convenient for the purpose of diagnosis of the disease. If the outer layers of cells of the midrib or the petiole are scraped away, a continuous brown region can be observed in necrotic bushes, where the phloem is discolored. In extreme cases, this necrosis may even spread to the main veins of the leaf. There are many cases where the bush looks perfectly normal but shows signs of necrosis, when examined by peeling the bark off.

Tree: Symptoms are mostly evident in tea recovering from pruning. This stands to reason as the virus in the bush is now manifest in fewer shoots, and therefore the virus is more concentrated in these few shoots, which in turn show marked symptoms. It is also possible that the multiplication of the virus is greater in the younger shoots.

Causal Organism: Phloem necrosis virus (Camellia Virus 1). This virus seems to have been observed only on the tea plant.

Transmission:

Neither the virus causing this condition nor the vector transmitting this virus has yet been identified. It has not been possible to mechanically transmit this disease either.

Epidemiology:

This disease was confined to higher elevations due to the fact that low temperature was conducive for the spread of this particular virus.

Integrated Disease Management:

1. As no cure for a virus disease in any crop has yet been discovered, the only practical method to control this disease is to destroy the virus by uprooting every tea bush that shows positive signs of the virus.
2. In the absence of a curative solution to the problem, the only remedy, as it now seems, is to uproot all affected bushes and re-supply them with healthy clonal plants.
3. Bushes afflicted by the virus could be identified at the time the field is recovering from pruning, and a further evaluation could be made at the end of the pruning cycle to determine the productivity of the bush.

KEYWORDS

- **tea crop disease**
- **etiology**
- **symptomatology**
- **management**

REFERENCES

Agnihothrudu, V. (1964). A world list of fungi reported on tea (*Camellia* spp.). *Annual Scientific Report*, Tocklai Experiment Station, India, Indian Tea Association for 1963: 64 (*Hypoxylon nummularium* inter alia; 3142).

Ajay, D., Balamurugan, A. and Baby, U. I. (2009). Survival of *Exobasidium vexans*, the incitant of blister blight disease of tea during off season. *International Journal of Applied Agricultural Research*. **4(2):** 115–123.

Almada-Rui E., Martinez-Tellez, M. A., Hernandez-Alamos, M. M., Vallejo, S., Primo-Yufera, E. and Vargas-Arispuro, I. (2003). Fungicidal potential of methoxylated flavors from citrus for *in vitro* control of *Colletotrichum gloeosporioides*, a causal agent of anthracnose disease in tropical fruits. *Pest Management Science*. **59:** 1245–1249.

Anita, B., Selvaraj, N. and Mani, M. P. (2011). Management of blister and grey blight in tea in nilgiris with a combination of carbendazim and mancozeb. *Pestology*. **35.** 29–33.

Arulpragasam, P. V., Addaickan, S. and Kulatunga, S. M. (1987). An inexpensive and effective method for the control of red root disease of tea. *Sri Lanka Journal of Tea Science*. **56(1):** 5–11.

Baby, U. I. (2001). Diseases of tea and their management: A review. In: P. C. Trivedi (ed.), *Plant Pathology*. Jaipur: Pointer Publication, pp. 315–327.

Baby, U. I. (2002). An overview of blister blight disease of tea and its control. *Journal of Plantation Crops*. **30:** 1–12.

Baby, U. I., Balasubramanian, S., Ajay, D., Premkumar, R. (2004). Effect of ergosterol biosynthesis inhibitors on blister blight disease, the tea plant and quality of made tea. *Crop Protection*. **23:** 795–800.

Baby, U. I., Ravichandran, R., Ganesan, V., Parthiban, R. and Sukumar, S. (1998). Effect of blister blight disease on the biochemical and quality constituents of green leaf and CTC tea. *Tropical Agriculture*. **75(4):** 452–456.

Balamurugan, A., Jayanthi, R., Muthukannan, P., Sanmugapriyan, R., Kuberan, T. and Premkumar, R. (2013). Integrated nutrient management by using bioinoculants in seedlings of tea (*Camellia sinensis*) under nursery. *International Journal of Advancements in Research & Technology*. **2(12):** 245–256.

Barman, H., Roy, A. and Das, K. S. (2015). Evaluation of plant products and antagonistic microbes against grey blight (*Pestalotiopsis theae),* a devastating pathogen of tea. *African Journal of Microbiology Research*. **9(18):** 1263–1267.

Bhujel, A. Singh, M., Choubey, M. and Singh, M. (2016). Pest and diseases management in Darjeeling tea. *International Journal of Agricultural Science and Research*. **6(3):** 469–472.

Chapman, R. L. (1984). An assessment of the current state of our knowledge of the Trentepohliaceae. In: D. Irvine and D. John (eds.), *Systematics of the Green Algae*. London: Academic Press, pp. 233–250.

Chapman, R. L. and Henk, M. C. (1985). Observations on the habit, morphology, and ultrastructure of *Cephaleuros parasiticus* (Chlorophyta) and a comparison with *C. virescens*. *Journal of Phycology*. **21:** 513–522.

Chen, Z. M. and Chen, X. (1990). *The Diagnosis of Tea Diseases and Their Control*. Shanghai: Shanghai Scientific and Technical Publishers.

De Weille, G. A. (1959a). The adoption of tea cultivation to the occurrence of blister blight. *Archaeology of Voor Thee Culture*. **20:** 161–192.

De Weille, G. A. (1959b). The fungus causing blister blight. *Archives of Tea Cultivation.* **20 (1):** 9–31.

Dutta, B. K. and Barthakur, B. K. (1998). *Recent Trends in Tea Disease Management Tea Research Association.* Calcutta: Tocklai.

Espinoza, J. G., Briceno, E. X., Keith, L. M., Latorre, B. A. (2008). Canker and twig dieback of blueberry caused by *Pestalotiopsis* spp. and a *Truncatella* sp. in Chile. *Plant Disease.* **92:** 1407–1414.

Fuchs, H. J. (1989). Tea environments and yield in Sri Lanka. *Tropical Agriculture*, Vol. 5. Weikersheim: Margraf Scientific Publisher, pp. 320.

Hainsworth, E. (1952). *Tea Pests and Diseases and Their Control.* Cambridge: Heffer.

Hasan, R., Rahman, M. H., Hussain, A., Muqi, A., Hossain, A., Ali, M. and Islam, M. S. (2014). Influence of topography, plant age and shading on red rust (*Cephaleuros parasiticus* karst.) disease of tea in sylhet region. *Journal of the Sylhet Agricultural University.* **1(2):** 227–230.

Holliday, P. (1980). *Fungus Diseases of Tropical Crops.* Cambridge, UK: Cambridge University Press, p. 607.

Horikawa, T. (1986). Yield loss of new tea shoots due to grey blight caused by *Pestalotia longiseta* Spegazzini. *Bulletin of Shizuoka Tea Experiment Station.* **12:** 1–8.

Huq, M., Ali, M. and Islam, M. S. (2007). Red rust disease of tea and its management. *Memorandum No.1*, BTRI, pp. 1–8.

Huq, M., Ali, M. and Islam, M. S. (2010). Efficacy of muriate of potash and foliar spray with fungtcides to control red rust disease (*Cephaleurous parasiticus*) of tea. *Bangladesh Journal of Agricultural Research.* **35(2):** 273–277.

Joshi, S. D., Sanjay, R., Baby, U. I. and Mandal, A. K. A. (2009). Molecular characterization of *Pestalotiopsis* spp. associated with tea (*Camellia sinensis*) in southern India using RAPD and ISSR markers. *Indian Journal of Biotechnology.* **8(4):** 377–383.

Joubert, J. J. and Rijkenberg, F. (1971). Parasitic green algae. *Annual Review of Phytopathology.* **9:** 45–64.

Keith, L., Ko, W. H. and Sato, D. M. (2006). Identification guide for diseases of tea (*Camellia sinensis*). *Plant Disease (PD)-33.* Manowa, HI: Cooperative Extension Service, University of Hawaii at Manowa, USA, pp. 1–4.

Kumhar, K. C., Babu, A., Bordoloi, M., Benarjee, P. and Rajbongsh, H. (2016). Comparative bioefficacy of fungicides and *Trichoderma* spp. against *Pestalotiopsis theae*, causing grey blight in tea (*Camellia* sp.): An *in vitro* study. *International Journal of Current Research in Biosciences and Plant Biology.* **3(4):** 20–27.

Lehmann-Danzinger, H. (2000). Diseases and pests of tea: Overview and possibilities of integrated pest and disease management. *Journal of Agriculture in the Tropics and Subtropics.* **101:** 13–38.

Liau, T. L. (1966). Blister blight and its control. *Taiwan Agriculture.* **11:** 1–5.

Madar, Z., Solel, Z. and Kimchi, M., (1991). Pestalotiopsis canker of Cypress in Israel. *Phytoparasitica.* **19(1):** 79–81.

Mann, H. H. (1901). Red rust: A serious blight of the tea plant. *Institut du transport aérien Bulletin.* **1:** 16.

Mann, H. H. (1906). Blister blight of tea. *Bulletin of Indian Tea Association.* **3:** 13.

Mann, H. H. and Hutchinson, C. H. (1904). Red rust. *Institut du transport aérien Bulletin.* **4:** 16.

Massee, G. (1898). Tea blights. *Kew Bulletin.* **1898:** 105–112.

McQuilken, M. P. and Hopkins, K. E. (2004). Biology and integrated control of *Pestalotiopsis* on container grown ericaceous crops. *Pest Management Science.* **60:** 135–142.

Me Rae, W. (1910). The outbreak of blister blight on tea in the Darjeeling district in 1908-1909. *Agricultural Journal of India*. **5:** 126–137.

Natesan, S. (1989). Physico-chemical studies on soils of the tea growing areas in south India. *Planters' Chronicle*. **84 (2):** 41–49.

Peal, S. E. (1868). Blister blight. *Journal of Agri-Horticulture Society of India.* **1:** 126.

Petch, T. (1923). *Diseases of Tea Bosh*. London: Macmillan, pp. xii + 220.

Ponmurugan, P., Baby, U. I. and Gopi, C. (2006). Efficacy of certain fungicides against *Phomopsis theae* under *in vitro* conditions. *African Journal of Biotechnology*. **5(5):** 434–436.

Ponmurugan, P., Saravanan, D. and Ramya, M. (2010). Culture and biochemical analysis of a tea algal pathogen, *Cephaleuros Parasiticus*. *Journal of Phycology*. **46(5):** 1017–1023.

Premkumar, R., Nepolean, P., Pallavi, R. V., Balamurugan, A. and Jayanthi, R. (2012). Integrated disease management of grey blight in tea. *Two and a Bud*. **59:** 27–30.

Ram, V. C. S. (1973). *Annual Report of the UPASI Scientific Department 1972-73*. Cinchona, Coimbatore: UPASI, pp. 7–27.

Ram, V. C. S. (1983). Pathogens and pests of tea. *Exotic Plant Quarantine Pests and Procedures for Introduction of Plant Materials*. Coonoor: UPASI Scientific Department, pp. 117–144.

Ram, V. C. S. and Mouli, B. C. (1983). Interaction of dosage, spray interval and fungicide action in blister blight disease control in tea. *Crop Protection*. **2:** 27–36.

Rattan, P. S. (1993). Incidence of *Phomopsis* stem and branch canker in Zimbabwe. *Tea Research Foundation Quarterly Newsletter*. **112:** 23–24.

Sana, D. L. (1989). *Tea Science*. Dhaka: Ashrafia Boi Ghar, pp. 224–226.

Sanjay, R. and Baby, U. I. (2005). Grey blight disease in tea. *Planter's Chronicle.* **101:** 4–9.

Satynanarayana, G. and Barua, G. C. S. (1983). Leaf and stem disease of tea in N-E. India with reference to recent advances in control. *Journal of Plantation Crops*. **45:** 34–56.

Sharma, K. C. (1960a). Red rust *Cephaleuros parasiticus* Karst. Disease of tea and associated crops in North-East India. *Memorandum No. 26*, 44–46.

Sharma, K. C. (1960b). Diseases of tea and associated crops in India. Indian Tea Association, *Memoir*. **26:** 34–67.

Sowndhararajan, K., Marimuthu, S. and Manian, S. (2013b). Integrated control of blister blight disease in tea using the biocontrol agent *Ochrobactrum anthropi* strain BMO-111 with chemical fungicides. *Journal of Applied Microbiology*. **114:** 1491–1499.

Sowndhararajan, K., Marimuthu, S. and Manian, S. (2013a). Biocontrol potential of phylloplane bacterium *Ochrobactrum anthropi* BMO-111 against blister blight disease of tea. *Journal of Applied Microbiology*. **114(1):** 209–218.

Sugha, S. K. (1997). Perpetuation and seasonal build-up of *Exobasidium vexans*, causal agent of blister blight of tea in Himachal Pradesh. *Tropical Science*. **37:** 123–128.

Surendramohan, M. (1995). Fertility status of the soil. *Planters' Chronicle*. **90(6):** 239–249.

Thakur, B. R. and Masand, S. (2015). Efficacy of Nativo 75 WG against blister blight in tea [*Camellia sinensis* (L.) O. Kuntze]. *Himachal Journal of Agricultural Research*. **41(1):** 86–88.

Venkataram, C. S. (1970). Milestones in blister blight control. UPASI *Tea Science Department Bulletin*. **28:** 53–63.

Venkataram, C. S. (1974). Blister blight of tea. In: S. P. Raychaudhuri, A. Varma, K. S. Bhargava and Mehrotra (eds.), *Advances in Mycology and Plant Pathology.* New Delhi: Professor R.N. Tandon's Birthday Celebration Committee, pp. 211–221.

PART II

Medicinal Crops

CHAPTER 7

DISEASES OF ISABGOL (*PLANTAGO OVATA* FORSK.) AND THEIR MANAGEMENT

N. M. GOHEL[1*], and B. K. PRAJAPATI[2]

[1]*Department of Plant Pathology, B. A. College of Agriculture, Anand Agricultural University, Anand, Gujarat 388110, India*

[2]*Directorate of Research, S. D. Agricultural University, Sardarkrushinagar, Gujarat 385506, India*

[*]*Corresponding author. E-mail: nareshgohel@aau.in*

ABSTRACT

Blond psyllium (Plantago ovata Forssk.) also known as "isabgol", is an annual herb with narrow linear rosette like leaves belonging to the family Plantaginaceae. Among 200 species of blond psyllium, Plantago ovata Forssk is known for superior quality of husk. Isabgol seeds and husk is used in medicines especially for relieving constipation. Isabgol is an important cash crop cultivated for its export and being of important medicinal value is reported to have larger demands and is traded in major medicinal drug markets of the world.

Some of the important diseases encountered in betelvine plantation are discussed in this chapter.

7.1 INTRODUCTION

Blond psyllium (*Plantago ovata* Forssk.) commonly known as "Isabgol" is an annual herb with narrow linear rosette like leaves belonging to the family *Plantaginaceae.* Among 200 species of blond psyllium, *Plantago ovata*

Forssk. is known for a good quality of husk. Isabgol seeds and husks are used in medicines for constipation. Isabgol is an important cash crop cultivated for its export and of important medicinal value. India commands monopoly in production and export of the seed and husk to the world market. In India, this crop is mainly grown as a commercial crop in Gujarat, Rajasthan, and Madhya Pradesh (Rastogi and Mehrotra, 1993). The husk has the property of absorbing and retaining water, and therefore it works as an antidiarrhea drug.

7.2 ECONOMIC IMPORTANCE OF ISABGOL CROPS

Husk of seeds contains colloidal mucilage or polysaccharide fractions mainly consisting of xylose, arabinose, and galacturonic acid with rhamnose and galactose. It is also used in modern food industries for the preparation of ice cream, candy, etc. (Jat et al., 2015). The crop is grown during winter months and commercially cultivated in North Gujarat, Rajasthan, Haryana, Madhya Pradesh, and Punjab (Rathore and Pathak, 2002a).

This crop is attacked by various bacterial and fungal diseases among which fungal diseases are prevailing. It has been found that downy-mildew-infected plants are more vulnerable to be attacked by *Alternaria alternata*. It causes considerable damage every year and sometimes become very severe which results in the loss in total yield.

7.3 MAJOR DISEASES OF ISABGOL CROPS

According to Mandal (2010), major diseases of significance are as follows:

1. Downy mildew: *Peronospora plantaginis* Underwood, *Peronospora alta* Fuckel and *Pseudopernospora plantaginis* Underwood
2. Damping-off of seedlings: *Pythium ultimum* (*P. ultimum*) Trow
3. Wilt disease: *Fusarium solani*.

7.3.1 DOWNY MILDEW

Introduction:

It causes extensive damage to the crop and makes the cultivation of isabgol crop fruitless (Rathore and Pathak, 2002b). Downy mildew causes considerable reduction in seed yield and yield attributes of isabgol.

Symptoms:

1. Downy mildew (*Peronospora plantaginis* Underwood) is one of the major biotic stresses experienced by this host plant.
2. Normally, this disease appears at the time of spike initiation.
3. Disease's symptoms mostly appear on leaves as chlorotic patches accompany with ashy-white downy growth during favorable environmental conditions.
4. It infects foliar as well as floral parts (Mandal and Geetha, 2001) and is one of the major constraints in successful cultivation of the crop.

Casual Organism:

It is reported to be caused by *Peronospora alta* Fuckel (Kapoor and Chowdhary, 1976), *Pseudoperonospora plantaginis* Underwood (Sharma and Pushpendra, 1998), and *Peronospora plantaginis* Underwood (Desai and Desai, 1969).

Pathogen:

Mycelium of the *Peronospora alta* is intercellular. Haustoria are formed within the tissues. Sporangiophores are slender, dichotomously branched, tapering, and curved at an angle less than the acute angle. Mycelium of *Peronospora plantaginis* is intercellular. Sporangiophores are slender, tree like, with the characteristic of an erect trunk. They are dichotomously branched and gray to pale yellow in color. They usually arise singly or in clusters from the stomata on the lower surfaces of leaves. Sporangia formed on the sporangiophores are subhyaline in color and are broadly elliptical to subglobose in shape (Patel et al., 2014).

Scientific Classification (Kirk et al., 2008)

1. Kingdom: Chromista (Straminopila)
2. Phylum: Oomycota
3. Class: Oomycetes
4. Subclass: Peronosporomycetidae
5. Order: Perenosporales
6. Family: Perenosporaceae
7. Genus: *Perenospora*
8. Species: *Parasemia plantaginis*

Disease Cycle:

The disease is both seed- and soil-borne in nature.

Epidemiology:

1. Early sowing, higher seed rate, higher dose of nitrogen, and frequent irrigations make the crop more susceptible to this disease.
2. The disease spreads rapidly in a humid weather or when the dewfall is high.

Management:

It is very difficult to keep the disease under control once it has affected the crop severely. Downy mildew is the major disease and causes severe yield loss, if not controlled. Hence, prophylactic measures should be adopted whenever the weather becomes cloudy and humid.

1. Sowing of isabgol in the end of November is recommended for the effective and economical management of downy mildew in isabgol.
2. In the absence of a commercial downy-mildew-resistant cultivar, disease can effectively be controlled by
   - o seed treatment with metalaxyl at the rate of 5 g/kg seed and
   - o spraying of metalaxyl mancozeb (MZ) (0.2%) together at 10-day intervals.
3. Seed treatment and foliar sprays of metalaxyl MZ were found effective in suppressing downy mildew disease's severity. Treatment of seeds with metalyxyl MZ and one foliar spray of the same followed by two sprays of MZ produced maximum net return (Mandal et al., 2007).
4. Seed treatment with metalaxyl (3 g/kg seed) followed by three sprays of metalaxyl MZ (0.1%) or fosetyl-Al (0.2%) or copper oxychloride (0.2%) at 10-15 days of intervals initiating from the appearance of the disease (Patel and Parmar, 2013).
5. For the effective and economical management of downy mildew in isabgol, spray metalaxyl 0.1% (I spray) and MZ 0.2% (II spray) at the 15-day interval, starting from the initiation of the disease.

7.3.2 DAMPING-OFF SEEDLINGS

Introduction:

Damping off incited by *Pythium* spp. in different crops is responsible for losses of multibillion dollar worldwide including India. Damping off is responsible for poor germination and stand of seedlings in nursery beds, and it is a common problem in the isabgol crop. Many workers reported severe losses caused by this disease not only in isabgol crops but also in others.

Symptoms:

1. Both pre- and postemergence damping off are noticed, with older plants being less susceptible than young seedlings (Chastangner et al., 1979).
2. Young seedlings are killed even before they emerge out of the soil, leading to pre-emergence damping off.
3. In the case of postemergence of damping off, the initial infection seen as a slightly darkened, water-soaked spots on the collar region of the growing seedlings.
4. The infection spreads very fast, and the seedlings are killed (Chastangner et al., 1979).

Causal Organism:

The sporangia of *P. ultimum* are terminal, spherical, 21.8 μm in diameter, and do not germinate to produce zoospores. Oogonia are terminal, smooth, and spherical with an average diameter of 18.3 μm. Antheridia are usually 1 per oogonium and arise immediately below the oogonium. Oospores are aplerotic, single, and spherical, averaging 16.3 μm in diameter with a smooth, thick wall (Chastangner et al., 1979).

Scientific Classification (Kirk et al., 2008)

1. Kingdom: Chromista (Straminopila)
2. Phylum: Oomycota
3. Class: Oomycetes
4. Subclass: Peronosporomycetidae
5. Order: *Pythiales*
6. Family: *Pythiaceae*
7. Genus: *Pythium*
8. Species: *P. ultimum*

Disease Cycle:

Pythium species is both a weak saprophyte and poor parasite. It is a soil-borne pathogen. The pathogen survives in soil through its survival structures, that is, oospores and sporangia present in plant debris or most commonly through its mycelium (Rangaswami, 1979). At the mycelial stage, the pathogen infects the host plant and multiplies very rapidly. Oospores take nutrients from the moist or irrigated soil to germinate and produce zoospores which disseminate rapidly in the presence of high soil moisture and ultimately attack the seeds and germinate.

Epidemiology:

1. The disease is serious in low-lying areas as well as in high-temperature regions with high relative humidity at the time of planting (Anonymous, 1979).
2. *P. ultimum* Trow is the predominant causal agent of the disease, although *Fusarium oxysporum* Schlecht is also known to cause damping off in 120-150-day-old plants (Russell, 1975).

Management:

1. Thiram or captan @ 2.5 g/kg seeds and the same fungicides may be used for drenching the seedlings during the postemergence period (Naik and Sinha, 1995).
2. Wherever the soil is sick, the damping off can also be managed by the solarization of such soil using polythene sheets of 500-750 gauges for six weeks during summer (Naik et al., 1997).

7.3.3 WILT DISEASE

Introduction:

Fusarium wilt is the most destructive soil-borne disease throughout the world. This pathogen blocks the xylem transport system, resulting in severe wilt and death of brinjal plants.

Symptoms:

Discoloration of roots of affected plants and accumulation of a large number of hyphae within the xylem vessels are the important characteristic of the disease. Pathogen caused browning of the vascular system. As a result of the disease, green leaves change into silver color. In severe attack, entire plants are wilted, and cortical root rots are observed.

Causal Organisms: *Fusarium solani* and *Fusarium oxysporum.*

Disease Cycle: The disease is soil borne in nature.

Management:

1. Field sanitation and proper crop rotation should be followed.
2. Cultural operation like deep summer ploughing and crop rotation with nonhosts will help to manage the disease.
3. Seed treatment with any systemic fungicide like Benlate or Bavistin @ 2.5 g/kg of the seed should be done.
4. Seed treatment with 4 g *Trichoderma viride* or 2 g carbendazim/kg seed is effective.
5. Drenching with 1% Bordeaux mixture or blue copper @ 0.25% may give protection.

7.4 SOME OTHER IMPORTANT DISEASES

7.4.1 POWDERY MILDEW

Symptoms:

The disease usually appears at the time of flowering. It is characterized by the appearance of the powdery mass of spores on the leaves. Such spots are small, white, or grayish in color. These spots enlarge gradually, and eventually cover the entire plant surface in due course.

Causal Organisms: *Erysiphe polygoni.*

Management:

Spraying of the crop with any wettable sulfur compound or karathane (0.2%) should be done after the appearance of the disease. Two or three sprayings at 15-day intervals may be required for management.

7.4.2 ALTERNARIA LEAF BLIGHT

Symptoms:

Older and matured leaves are more commonly affected by this disease. First of all, dull green to yellowish-colored lesions are observed on the leaf tips. In severe cases, irregular necrotic spots are formed on the leaves of the affected plants. These spots enlarge gradually, resulting in the blackening and drying of the affected leaves. Shriveling and blackening of seeds of infected plants are also observed.

Causal Organisms: *Alternaria alternata* (Fr.) Keissler.

Management:

1. Spray suitable fungicide as used in different leaf spots.

7.5 OTHER DISEASES OCCURRING ON ISABGOL

Mandal (2010) also reported other diseases occurring on Isabgol; one of them is leaf spot disease that is caused by *Septoria plantaginae*.

KEYWORDS

- **isabgol crop disease**
- **etiology**
- **symptomatology**
- **management**

REFERENCES

Anonymous. (1979). Isaptent: A new cervical dilator. *CDRI Annual Report*. p. 12.

Chastangner, G. A., Orawa, J. M. and Sammetan, K. P. V. (1979). Agrochemical evaluation of genetic resources of some medicinal plants. *Indian Drugs*, **17**(2): 411-415.

Desai, M. V. and Desai, D. B. (1969). Control of downy mildew of isabgol by aureofungin. *Hind. Antibiot. Bull.*, **11**(4): 254-257.

Jat, R. S., Reddy, R. N., Bansal, S. and Manivel, P. (2015). *Good Agricultural Practices for Isabgol: Extension Bulletin*. Boriavi, Anand: ICAR, Directorate of Medicinal and Aromatic Plants Research.

Kapoor, J. N. and Chowdhary, P. N. (1976). Note on Indian microfungi. *Indian Phytopath.*, **29**: 348-352.

Kirk, P. M., Cannon, P. F., Minter, D. W. and Stalpers, J. A. (2008). *Ainsworth and Bisby's Dictionary of the Fungi* (10th Ed.). Oxon, UK: CAB International.

Mandal, K. (2010). Disease of some important medicinal crops and their management. *Microbial Diversity and Plant Disease Management*. Germany: VDM Verlag Dr. Muller, p. 509.

Mandal, K., Gajbhiye, N. A. and Maiti, S. (2007). Fungicidal management of downy mildew of isabgol (*Plantago ovata*) simulating farmers' field conditions. *Australasian Plant Pathology*, **36**: 186-190.

Mandal, K. and Geetha, K. A. (2001). Floral infection of downy mildew of isabgol. *J. Mycol. Plant Pathol.*, **31**: 355-357.

Naik, M. K., Reddy, M. V., Sinha, P. and McDonald, D. (1997). Differential solarization – a decision making tool for management of soil borne plant pathogens. *Annual Meeting and Symposium of Indian Phytopathological Society held at UAS, Bangalore*.

Naik, M. K. and Sinha, P. (1995). Control of nursery diseases of vegetables. *National Symposium on Plant Protection and Environment held by the Plant Protection Association, Madras*.

Patel, N. N. and Parmar, R. G. (2013). Effects of fungicides, plant extracts and bio agent on downy mildew of isabgol (*Plantago ovata* Forsk.). *Int. J. Plant Protect.*, **6**(1): 142-144.

Patel, N. N., Parmar, R. G. and Patel, S. T. (2014). Varietal screening of isabgol (*Plantago ovata* Forsk.) against downy mildew. *Int. J. Plant Protect.*, **7**(1): 243-245.

Rangaswami, G. (1979). *Diseases of Crop Plant in India* (2nd Ed). New Delhi: Prentice Hall of Indian Pvt. Ltd, pp. 298-302.

Rastogi, R. P. and Mehrotra, B. N. (1993). *Compendium of Indian Medicinal Plants*, Vol. III. New Delhi, India: Central Drug Research Institute, Lucknow and Publications & Information Directorate.

Rathore, B. S. and Pathak, V. N. (2002a). Studies on the tissue culture and host range of *Peronospora alta* causing downy mildew of blond psyllium. *J. Mycol. Pl. Pathol.*, **32**(2): 201-203.

Rathore, B. S. and Pathak, V. N. (2002b). Effect of seed treatment on downy mildew of blond psyllium. *J. Mycol. Pl. Pathol.*, **32**(1): 35-37.

Russell, T. E. (1975). *Plantago* wilt. *Phytopathology*, **65**: 359-360.

Sharma, M. P. and Pushpendra. (1998). A new pathogen causing downy mildew of isabgol (*Plantago ovata* Frosk.) (Abstr.). *J. Mycol. PI. Pathol.*, **28**: 74.

CHAPTER 8

DISEASES OF SENNA (*CASSIA ANGUSTIFOLIA* M. VAHL.) AND THEIR MANAGEMENT

B. K. PRAJAPATI[1*] and N. M. GOHEL[2]

[1]*Directorate of Research, S. D. Agricultural University, S. D. Agricultural University, Sardarkrushinagar, Gujarat 385506, India*

[2]*Department of Plant Pathology, B. A. College of Agriculture, Anand Agricultural University, Anand, Gujarat 388110, India*

[*]*Corresponding author. E-mail: bindesh_prajapati@yahoo.in.*

ABSTRACT:

Senna crop plant parts were used by physicians from early civilization. The plant is mainly cultivated in different countries including Sudan, Egypt, and India. In India, it is cultivated in southern districts of Tamil Nadu. In medicine, sennosides being laxative in action are used as calcium sennoside tablets. India is the major producer and exporter of senna and there is enough scope for the expansion of area under cultivation of this crop to earn more foreign exchange while catering the needs of the world demand. The crop is known to suffer from diseases like damping off, leaf spots, root rot, root knot, die back, and leaf blight. Among these diseases, leaf blight caused by *Alternaria alternata* is the most serious disease causes lots of damage to the crop. As a result of leaf blight infection, defoliation occurs that affect the crop yield very badly. Use of chemicals for the management of senna blight may affect its medicinal value.

8.1 INTRODUCTION

The plant is mainly cultivated in different countries including Sudan, Egypt, and India. The leaves and green, immature pods contain glycocides like sennocides A and B, which are extensively used as a laxative particularly for habitual constipation. In medicine, sennosides being laxative in action are used as calcium sennoside tablets. The crop can be grown as an irrigated crop throughout Andhra Pradesh on marginal red laterites, red soils, coarse gravelly soils, alluvial loams and on rich alayey soils, ranging in pH from 7.0 to 8.5 (Anon., 2015).

Senna was used by the physicians from early civilization. The plant is mainly cultivated in different countries including Sudan, Egypt, and India. In India it is cultivated in southern districts of Tamil Nadu. In medicine, sennosides being laxative in action are used as calcium sennoside tablets.

8.2 DISEASES OF SENNA (CASSIA ANGUSTIFOLIA M. VAHL.)

Leaf spot/blight: *Alternaria alternata* (*A. alternata*) (Fr.) Keissler

Alternaria cassiae Jurair and Khan

Phyllosticta sp.

Cercospora sp.

Brown pod rot: *Bipolaris australiensis*

Damping off: *Pythium* sp., *Rhizoctonia bataticola.*

8.3 DAMPING OFF

Symptoms:

Affected seedlings first show a slight yellowish patch just above the ground level. This darkens with time, and the tissues soften and the whole seedling coalesces to the ground. Young seedlings are extremely susceptible to the disease, reducing the plant population drastically.

Causal Organism: *Rhizoctonia bataticola.*

Scientific Classification (Kirk et al., 2008)

1. Phylum: Basidiomycota
2. Class: Agaricomycetes
3. Order: Cantharellales
4. Family: Ceratobasidiaceae
5. Genus: *Rhizoctonia*
6. Species: *bataticola.*

Disease Cycle and Epidemiology:

The fungus overwinters in the soil and crop residue and is the primary source of inoculum. The disease development favored at high humidity and temperature ranging from 30 °C to 37 °C. The disease spread fast if the fields are affected by stagnating water.

Management:

Seed treatment with thiram or captan at 2.5 g/kg of seeds (Anonymous, 2011).

8.4 COLLAR ROT DISEASE

Symptoms:

The infected plants usually show the characteristic of blackening on the collar region extending upward, leading to the death of the seedling.

Causal Organism: *Macrophomina phaseolina* (Tassi) Gold (Syn. *Rhizoctonia hataticola* (Taub) Butler).

Sclerotia of *Rhizoctonia bataticola* within roots are black, small, and hard, measuring 100 mm-1.0 mm in diameter. Pycnidia are dark brown, solitary, immerged, and become erumpent in due course. Pycnidia consist of multicellular walls that are heavily pigmented and open through apical ostioles. Conidiophores are hyaline, short, obpyriform to cylindrical in shape. Conidia are hyaline and ellipsoid to obovoid in shape. They measure 14-30 × 5-10 mm in size.

Disease Cycle and Epidemiology:

The fungus is a soil-borne pathogen. High humidity and temperature ranging from 30 °C to 37 °C favor the disease development.

Management:

1. Follow the crop rotation.
2. Seed treatment before sowing should be done with thiram or captan @ 2.5 g/kg of the seed.
3. Drench the field with Bavistin (0.1%) or Brassicol (0.2%) as this has been reported effective.

8.5 LEAF SPOT/BLIGHT

Symptoms:

Alternaria leaf spot/blight disease is visible first as minute pale yellow spots on the leaf blade and the margins. As the spots grow in size, they become circular and irregular in shape with their coalesces, and turn dark brown to black in color. The lesions vary from a minute spot to about 8 mm in diameter. With the advancement of the disease, leaf tips and margins die, and the necrotic tissues increase in size. A few to many spots may coalesce to form large irregular patches, covering almost the entire leaflet. In severe infection, leaves start drying and drooping, resulting in a poor yield. Concentric rings are visible in advanced leaf spots. The green color of leaves changes from pale green and greenish yellow. The infected leaves drop off prematurely and thus affects yield considerably. In advanced stages, pods are also found affected with brownish-black spots. The content of sennosides in the leaves is inversely proportional to the intensity of the disease.

Causal Organism:

Senna has been reported to be affected by leaf blight caused by *A. alternata* (Fr.) Keissler (Saxena et al., 1981) in India.

A. alternata and *Alternaria tenuissima* were also found to cause foliar blight of *Cassia fistula* and *Cassia tora* in India and Pakistan (Lenne, 1990), while *A. alternata* was observed as the most common pathogen causing defoliation with severe yield losses (Patel and Pillai, 1979).

Studies on the cultural variability of *A. alternata*-infecting senna plants showed that all the six isolates were different in terms of colony characters, colony diameter, sporulation, size of conidia, and septation. The colony diameter ranged from 45 to 47 mm. In general, conidial length and width were found in between 37.55 and 51.60 μm and 13.60 and 19.30 μm, respectively. The numbers of horizontal and vertical septa varied between 4 and 8 and 2 and 4, respectively (Tetarwal et al., 2008).

Scientific Classification (Kirk et al., 2008)

1. Phylum: Ascomycota
2. Class: Dothidiomycetes
3. Order: Pleosporales
4. Family: Pleosporaceae
5. Genus: *Alternaria*

Disease Cycle and Epidemiology:

Cloudy days and humid weather conditions are conducive to the spread of the disease. Tetarwal and Rai (2007) studied the effect of epidemiological factors on the Alternaria blight of senna. They reported that a comparatively higher disease intensity was observed during the months of July and September in both the years. The correlation studies suggested that relative humidity was a key factor for the development of the disease. The pathogen survived only up to 8 months under natural conditions but up to 10 months *in vitro*. On inoculation, eight-week-old plants showed maximum leaf blight as compared to young/older plants. Six-to-eight-day-old culture of *A. alternata* showed maximum virulence.

Since the senna plant is perennial shrub producing green foliage throughout the year, the pathogen has advantage to complete the disease cycle on and around the plant without any break.

Management:

1. Seed treatment with thiram at 3 g/kg seeds is also beneficial. Two to three sprayings of Mancozeb 75 WP @ 0.15% at the fortnightly interval can be carried out to check the disease. Harvesting of leaves and pods must be done after 25-30 days after the spray of pesticides (Jat et al., 2015).
2. Tetarwal and Rai (2013) tested different neem products with various concentrations (1000, 2000, 5000, and 10,000 ppm) against *A. alternata*, the causal agent of Alternaria blight of senna, of which Nimin was found most effective in inhibiting the mycelial growth, spore germination of *A. alternata* and in disease control followed by neem gold and nimbicidine (Walker, 1982).
3. *Tinospora cordifolia* stem extract proved effective in reducing the Alternaria leaf blight of senna (Anonymous, 2013).

8.6 COLLETOTRICHUM LEAF SPOTS

Symptoms:

The disease appears in the form of elliptic or oblong spots of variable size. In the initial stages of infection, the spots are small and measure only few centimeters in length and are 2–4 cm in breadth, which later may increase in size. Further, two or more leaf spots coalesce and develop into irregular patches that occupy a major portion of the leaf, which eventually dries up.

Casual Organism: *Colletotrichum gloeosporioides* (*C. gloeosporioides*) (Penz) Sacc. (perfect stage—*Glomerella cingulata* (Stonem) Spauld).

Acervuli within the lesions axe characteristically glabrous, round, elongated, or irregular in shape. Setae are variable in length, 1–4 septate, and are about 200 × 4-8 mm in size. Setae appear brown, slightly swollen at the base, and tapering to the apex. Conidia are more variable in size and are hyaline, aseptate, uninucleate, cylindrical, or slightly ellipsoidal with a rounded apex and a narrow truncate base. Perithecia are occasionally formed in young cultures.

Scientific Classification (Kirk et al., 2008)

1. Phylum: Ascomycota
2. Class: Sordariomycetes
3. Order: Glomerellales
4. Family: Glomerellacaea
5. Genus: *Colletotrichum*

Disease Cycle and Epidemiology:

The fungus *C. gloeosporioides* survives in dead twigs and injured plant tissues and forms an abundance of acervuli and conidia. Conidia can spread over relatively short distances by rain splash or overhead irrigation. Ascospores are airborne and important in long-distance dispersal. Conidia that come in contact with leaves, twigs, and fruit germinate to produce appressoria and quiescent infections that result in tissue necrosis. This tissue is subsequently colonized, acervuli are formed, thus completing the pathogen's life cycle. Dead wood and plant debris are primary sources of inoculum.

Environmental conditions favoring the pathogen growth are temperature ranging 25 °C–28 °C being optimum, pH range of 5.8–6.5, and high humidity.

Management:

1. Three sprays of Mancozeb 75% WP (0.15%) at the interval of 15 days have been recommended.

8.7 BROWN POD ROT

Symptoms:

The disease initially appeared in the form of brown, circular to irregular, minute to large necrotic spots on the pods that later turn into pod rot. Similar symptoms are also observed on leaves and stem. Seeds in infected pods are malformed and have reduced viability (Saroj et al., 2011).

Casual Organism:

This disease is caused by *Bipolaris australiensis*. Culture was initially white, but after three days acquired an olive green to black cast found to contain cylindrical conidia with three pseudosepta. Sizes ranged from 13–35 μm to 8–10 μm. Conidiophores were brown, simple, branched, geniculate, and sympodial. The hilum was characterized by a flattening of the basal cell, slightly protruding at the tip (Fang et al., 2007).

Scientific Classification (Kirk et al., 2008)

1. Phylum: Ascomycota
2. Class: Dothidiomycetes
3. Order: Pleosporales
4. Family: Pleosporaceae
4. Genus: Bipolaris

Saroj et al. (2011) reported brown pod rot of senna caused by *Bipolaris australiensis* for the first time from Lucknow, India. The fungus initially produced silky-smooth, greyish-white colonies that later became olive green to black with a raised greyish periphery.

Management:

1. Seed treatment before sowing should be done with carbendazim @ 2.0 g/kg of seed.
2. Spraying with carbendazim (0.1%) has been reported effective.

KEYWORDS

- **senna crop disease**
- **etiology**
- **symptomatology**
- **management**

REFERENCES

Anonymous (2011). http://eagri.org/eagri50/HORT282/pdf/lec32.pdf

Anonymous. (2013). *Annual Report 2013-14*. Boriavi, Anand: Directorate of Medicinal and Aromatic Plants Research, pp. 70.

Anonymous. (2014). http://www.croppro.com.au/crop_disease_manual/ch11s04.php

Anonymous. (2015). http://www.ikisan.com/medicinal-plants-senna.html

Anonymous. (2018). https://en.wikipedia.org/wiki/Alternaria_alternata

Boyette, C. D. (1988). Biocontrol of three leguminous weed species with *Alternaria cassia*. *Weed Technology*, **2:** 414-417.

Fang, K. F., Huang, J. B. and Hsiang, T. (2007). First report of brown leaf spot caused by *Bipolaris australiensis* on *Cynodon* spp. in China. *Plant Pathology.* **56:** 349.

Jat, R. S., Reddy, R. N., Bansal, R. and Manivel, P. (2015). *Good Agricultural Practices for Senna (Extension Bulletin)*. Boriavi, Anand: ICAR, Directorate of Medicinal and Aromatic Plants Research.

Kirk, P. M., Cannon, P. F., Minter, D. W. and Stalpers, J. A. (2008). *Dictionary of the Fungi* (10th ed). Wallingford: CABI.

Lenne, J. M. (1990). Diseases of *Cassia* species: a review. *Tropical Grasslands*, **24:** 311-324.

Patel, K. D. and Pillai, S. N. (1979). Effect of leaf spot disease on sennoside content in senna leaves. *Indian Drugs*, **17:** 1-2.

Saroj, A., Alam, M., Qamar, N., Khaliq, A. and Sattar, A. (2011). First report of *Bipolaris australiensis* causing pod rot of senna in India. *New Disease Reports*, **23:** 28.

Saxena, A. K., Jain, S. K. and Saksen, S. B. (1981). A note on new diseases caused by *Alternaria alternata*. *National Academy of Science Letters*, **4:** 267.

Tetarwal, M. L. and Rai, P. K. (2007). Effect of epidemiological factors on Alternaria blight of senna (*Cassia angustifolia* Vahl.). *Annals of Plant Protection Sciences*, **15(1):** 148-150.

Tetarwal, M. L. and Rai, P. K. (2013). Management of Alternaria blight of senna (*Cassia angustifolia* Vahl.) with neem (*Azardiracta indica*) products. *AGRES*, **2(1):** 78-82.

Tetarwal, M. L., Rai, P. K. and Shekhawat, K. S. (2008). Morphological and pathogenic variability of *Alternaria alternata* infecting senna (*Cassia angustifolia*). *Journal of Mycology and Plant Pathology*, **38(3):** 375-377.

Walker, H. L. (1982). Seedling blight of sickle pod caused by *Alternaria cassiae*. *Plant Disease*, **66(5):** 426-428.

PART III

Mushroom Diseases and Others

CHAPTER 9

DISEASES AND COMPETITOR MOLDS OF MUSHROOMS AND THEIR MANAGEMENT

MOHSINALI M. SAIYAD[1,*] and J. N. SRIVASTAVA[2]

[1]*College of Agriculture, Vaso, Anand Agricultural University, Anand, Gujarat 388110, India*

[2]*Department of Plant Pathology, Bihar Agricultural University, Sabour, Bhagalpur, Bihar, India*

[*]*Corresponding author. E-mail: munna11983@aau.in*

ABSTRACT

Mushrooms produce quality food if we compared it with food produced from green plants. Its versatile nature will supports it as a valuable tool in medicinal and metabolic researches. The edible mushroom not only provides food with rich in nutrients but also with plenty of medicinal benefits too. Similarly to crop plant, mushrooms are also subjected to different diseases during cultivation. Both primary and secondary factors may lead to different diseases during cropping. The major causes for bacterial and fungal diseases likely to come from using infected compost and substrates used for cultivation. Most of the viral diseases are likely to get spread from spores. Mushroom cultivation is severely affected from diseases in terms of losses and thus proper management is needed throughout the cropping period.

9.1 INTRODUCTION

Mushroom, also known as toadstool, puff balls, etc., is a saprophytic fungus mostly edible (sometimes poisonous or nonedible) in nature. It does not contain chlorophyll; hence, it cannot make its own food so derives food from other sources. Mushroom is a macro fungus, saprophytic, and heterotrophic in nature, having a fruiting body which can be seen by the naked eye. Naturally it grows as both epigeous and hypogeous.

The term "mushroom" is originally taken from a French word "mousseron" which means moss, foam, yeast, fungus, etc. It is an important crop of fungal origin that can be cultivated on various agricultural residues. There is a vast biodiversity of edible as well as nonedible mushroom species found and grown all over the world. A mushroom is a quality food that provides high-quality protein, vitamins, minerals, and nutrients that no other food can give. Its high medicinal value gives it a priority for medical research in the modern world.

Bonnefons (1650) was the pioneer researcher from Paris (France) who first cultivated mushroom in the field condition. Tourneforte (1707) first described the use of casting soil for the cultivation of mushroom. Around 1825, the cultivation of mushroom was started in caves in the Netherlands, and later that was followed in France during 1848 and successively in Europe. Callow (1831) showed that mushroom can be produced the whole year around in rooms in England. The first commercial cultivation of mushroom was started in France during the 1930s with the white button mushroom.

9.2 ECONOMIC IMPORTANCE OF MUSHROOM

Mushrooms produce better quality food if we compare them with food produced from green plants. Its versatile nature supports it as a valuable element in medicinal and metabolic researches. The edible mushroom not only provides food with rich nutrients but also with plenty of medicinal benefits.

9.3 NUTRITIONAL VALUES OF MUSHROOMS

Mushrooms are one of the unexplored treasures of nutrients provided by nature. In India, cereals are main staple food which are although rich in carbohydrates, are a very poor source for protein. Addition of mushroom in the Indian diet will definitely solve this problem as they are an ultimate source of high-quality protein. Irrespective of age (i.e., from child to adult), mushroom can be used as a food for everyone as a supplement of nutrients.

They are a rich source for vitamins, fibers, and minerals. Consumption of mushroom provides low calorie with little fat. It contains ergosterol and is devoid of cholesterol; hence, it is good for health. The nutrient aspect of mushroom varies with growth, environmental factors, as well as the nature of species. Sugars, amino acids, and starches found in mushrooms are fully digestible easily, while crude fibers of mushrooms are partially digestible in nature.

Mushroom is a very rich source of vitamin B complex, that is, $B_{2,}$ $B_{3,}$ B_5, and B_{12}. Vitamin B has a significant role in the functioning of the nervous system. It provides quality minerals such as selenium, copper, and potassium which have direct influence on heart, nervous system, and blood pressure, respectively. Selenium found in mushroom strengthens immunity and fertility in men, as well as it helps to fight against cancer. Edible mushrooms contain a large amount of polyunsaturated fatty acids. They also contain ergosterol which helps in the synthesis of vitamins in human being, that is, vitamin D. Upon direct exposure to sunlight, ergosterols are immediately converted to vitamin D_2 naturally.

From the nutritional point of view, mushroom contains a higher amount of protein. It contains amino acids in varied proportions, that is, a higher proportion of threonine and valine and a lower proportion of sulfur containing amino acids (i.e., methionine and cysteine). A very high water-to-fiber ratio (80%-90% to 10%-20%) makes mushroom an easily digestible food among all. Mushroom serves as an excellent source of vitamins such as vitamins B and C. It also provides a higher content of sodium, potassium, phosphorous, and other minerals. Mushroom can provide most of the essential minerals and nutrients except iron and calcium in a greater proportion. Mushroom has a naturally occurring antioxidant, that is, Ergothioneine which helps body to fight against stress condition and also has an anti-aging property. It also has a significant role in maintaining healthy eyes, liver functions, maintaining kidney health, and skin condition.

9.4 MEDICINAL VALUES OF MUSHROOM

Mushroom has exceptional medicinal properties. It not only restores human health but also helps to fight against lethal diseases. It strengthens an overall immunity in the human body as well as helps body in fighting against of noncurable diseases like cancer. It also helps in curing various diseases such as diabetes, control of blood sugar, tumors, and inflammation.

Mushroom has a very low content of fat, no cholesterol, and high content of fatty acids that are mainly unsaturated in nature. Due to such properties, it is an ideal food for healthy heart and also helps in fight against some cardiovascular diseases. It is also a rich source for potassium and has a very low content of sodium that helps body in maintaining blood circulation properly as well as maintaining salt balance within the body. Consuming mushrooms on a daily basis helps in reducing cholesterol levels in blood. As it does not contain cholesterol, it serves as an excellent food for patients with high blood pressure.

Mushroom has very low calorific value, fats, and sugars which make them an ideal food for diabetic patients. It also helps in losing body weight as quality proteins present in mushroom burns cholesterol found within the body.

Mushroom provides some quality compounds that help body in restricting tumor development. Most of the mushroom species, especially edible mushrooms, help in preventing cancers such as breast cancer, pancreatic cancer, and prostate. It also provides polysaccharides which helps in reducing side effects of chemotherapy in body. The mushroom-derived polysaccharides are helpful in preventing the free radicals' movement in the body and also help in mitigating the aging mechanism in human beings. The oligosaccharide derived from mushrooms acts as prebiotics and helps in maintaining healthy stomach as well as an overall digestion system.

Mushroom has quality polysaccharides (especially, beta glucans) and minerals which are responsible for strengthening and regulating the immune system.

9.5 MORPHOLOGY OF MUSHROOM

The mushroom body is divided into three basic structures, that is, the base, the tip, and the cap. However, sometimes certain parts are not formed. The sporophore of mushroom is of two main types, that is, ascocarp and basidiocarp. The ascocarp may be cup shaped, bearing a sponge-like structure, or bell shaped. In contrast, mushroom belongs to Basidiomycotina form basidiocarps which may be stipitate, substipitate, or sessile in shape.

The mycelium is formed underground that helps in absorbing water, food, and nutrients from soil. Depending upon the nutrients it gets, it may live for centuries. The sporophore which may sometimes be called fruit develops above ground. It has an umbrella-kind shape and dies quickly to yield other structural parts. The main aboveground part adjoining to the mycelium and on which the cap is formed is known as the stalk or stem. This part grows very quickly as compared to other structural parts as it absorbs a large amount of water and nutrients. Upon maturity, the stalk yields an umbrella-type shape. Spores are released by gills which are located at the inner side of the cap or pileus. Spores that are released from gills are transported via air and develop a new mycelium as soon as they reach a humid or wet area.

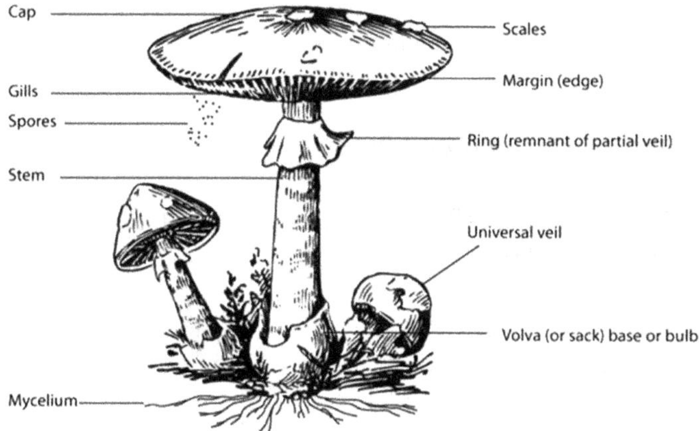

FIGURE 9.1 Morphology of mushroom.

A mushroom develops from a primordium, a nodule—a pinhead-like structure. Upon maturity, the primordium develops an egg-shape structure known as "button" which may be surrounded by a universal veil. As the egg enlarges, universal veil gives rise to another structure known as volva.

9.6 BIOLOGY OF MUSHROOM

Earlier, mushrooms were classified into the Kingdom *Plantae*, but as they resemble most of the characters nearest to fungi, they are now put under the Kingdom fungi. Most of the cultivated mushroom species belong to the phyla Basidiomycota and Ascomycota. Furthermore, mushrooms belong to the order Agaricales and class Agaricomycetes. A mushroom is heterotrophic in nature and has big sporophores in its structure. It can grow and survive throughout the year on varied habitats. Mushrooms are pleomorphic in nature and have different size, shape, taste, and color as per the species. Most of the mushroom species are edible, while some are nonedible or poisonous in nature (i.e., *Amanita* spp. and *Cortinarius* spp.).

9.7 BIODIVERSITY OF MUSHROOM

Out of thousands of mushroom species available in the world, most are naturally grown and some are cultivated. Although most of species have great nutritional and medicinal properties, some species are found as dangerous or lethal, too. Mushroom is divided into two broad categories based on consumption and edibility, that is, edible and nonedible/poisonous (Sharma, 1994a).

9.7.1 EDIBLE MUSHROOM SPECIES

9.7.1.1 BUTTON MUSHROOM/TEMPERATE MUSHROOM (AGARICUS BISPORUS)

It is cultivated all over the world due to its wide adaptability. Morphologically, it has a white creamy and large button-shaped soft fruit body with good flavor, taste, and nutrition. The cultivation of the button mushroom was started in France during the sixteenth century (Seth, 1977). In India, it is largely cultivated in Himachal Pradesh, Jammu and Kashmir, Haryana, and Uttar Pradesh. The important cultivated species of the button mushroom are *Agaricus bisporus* (*A. bisporus*) and *Agaricus bitorquis* which are known as "white button mushroom" and "temperate tolerant white button mushroom," respectively (Biswas et al., 2012).

9.7.1.2 OYSTER MUSHROOM (PLEUROTUS SPP.)

It is one of the most important mushroom species cultivated widely in China and South Korea. In India, it is largely cultivated in the southern and northeastern provinces. This mushroom species can be grown on any agricultural waste without any special care. However, it will grow best on rice straw or wheat straw. It has a high shelf life, pleasant flavor, and excellent nutritional value which make it a popular commercial mushroom in the world. Due to the oyster shape of pileus, it is commonly known as the oyster mushroom. In India, locally it is known as "dhingri mushroom." The important cultivated oyster mushroom species are *Pleurotus sajor-caju*, *Pleurotus flabellatus*, *Pleurotus ostreatus*, *Pleurotus florida*, *Pleurotus sapidus*, *Pleurotus eous*, and *Pleurotus eryngii* (Krishna, 1993).

9.7.1.3 MILKY MUSHROOM (COLOCYBE INDICA)

It is locally known as "dudh chhata" in West Bengal, India. It normally yields fruits in hot temperature so is grown in the summer season. It is naturally grown under the roadside trees or in forest. The spore produced by this mushroom has milky white color and a delicious flavor. It has a longer shelf life than those of other cultivated species. Due to a high fiber content, it is very good food in the conditions such as, hyperacidity and constipation. It is also used in food industries for the preparation of pickles. Structurally, it has white- or pale-colored sporophore, a flattened pileus of white color, a regular and smooth margin, and distinctly formed gills which are crowded, separable, and having white color and thick appearance.

9.7.1.4 SHIITAKE MUSHROOM (LENTINUS/LENTINULA EDODES)

It is one of the widely grown mushrooms in East Asia, particularly in Japan, China, South Korea, and Taiwan. It usually grows on trees such as Oak, Horn bean, and Chestnut. Pileus is pale to dark reddish brown, deep brown toward the center, at first convex and later depressed, and the margin regular with persistent veil remnants. Gills initially have white color which changes to brown as the mushroom matures. It contains lentinan, the water-soluble polysaccharide, which helps in reducing plasma cholesterol in human, and controls blood pressure. It also has anticancer, anti-HIV, and antidiabetic properties.

9.7.1.5 BLACK EAR MUSHROOM (AURICULARIA SPP.)

It is locally known as "Jew's ear," "wood ear," or "Mou-erh" (in Chinese). It has typical gelatinous, irregular, and foliose fruit bodies cultivated on dead woods. Sporophore of *Auricularia auricula* has short stem, ear shaped, yellow brown color when fresh, and olive brown when dry. The major cultivated species includes *Auricularia aricula and Auricularia polyrich*, respectively (Garcha, 1984).

9.7.1.6 PADDY STRAW MUSHROOM (VOLVARIELLA VOLVACEA)

The paddy straw mushroom is used extensively in Asian cuisines and widely cultivated throughout East and Southeast Asia. It prefers subtropical climates with high annual rainfall for better growth. It is mostly grown on paddy straw. It contains good combinations of all attributes such as flavor, aroma, delicacy, high-quality protein, vitamins, and minerals. The fruiting body develops from an advanced hyphal aggregate called as primordia.

The developmental process has various structural stages called "button," "eggs," "elongation," "mature," respectively. The differentiation process starts at the "button" stage which finally forms an umbrella-shaped fruit bodies upon maturity (Bhavani Devi and Nair, 1986; Kannaiyan and Prasad, 1978).

9.8 NONEDIBLE/POISONOUS MUSHROOMS

The mushroom toxin produces lethal damage to human body upon ingestion, that is, mycetism. Most of the poisonous mushrooms are classified mainly into two genera, namely *Amanita* and *Cortinarius*. Some other genus also has moderate to lethal toxin, that is, *Galerina*, *Lepiota*, and *Conocybe*.

9.8.1 AMANITA SPP.

The majority of poisonous mushroom species belong to this genus. Not every *Amanita* species is dangerous, but some may kill, too. The *Amanita phalloides* (*A. phalloides*) is the most poisonous species of this genus. It is also known as "toadstools" and "death cap." Among the total causalities resulted from the poisonous effect of mushroom, this genus has almost accounted for 90%-95% of causalities. It produces "α-amanitin" toxin which has lethal and destructing effect on liver and kidneys. The death cap is widely distributed throughout Europe. It mainly appears in the summer and autumn seasons. Morphologically, it has a green-color cap and stipe, and white-color gills. From the historical point of view, in the deaths of Roman Emperors in the seventeenth century, the targeted mushroom was of genus *Amanita*.

Destroying angels (*Amanita virosa*) also have poisonous properties. Morphologically, they have white stalk and free gills (not attached to stalk), whitish yellow cap, and a veil. The presence of the universal veil or a volva is one of the typical characters of this species. It prefers to grow in a small group.

The fool's mushroom (*Amanita verna*), a close relative of *A. phalloides*, is a well-known species, mostly found in Europe. Due to its nature, sometimes it is called "death angel." It is mainly grown in summer and autumn on deciduous trees. It produces white-colored pileus and stipes. It has a large, bag-like volva with a membranous annulus which is white in color. The spores produced are elliptical and smooth.

9.8.2 *CORTINARIUS SPP.*

Cortinarius mushrooms, containing over 2000 different species, are found worldwide. During the younger stage of life, this mushroom has a cortina (veil) between the stalk and the pileus, hence the name curtained. It produces spores which are rusty brown in color and have very lethal effect on the digestive system. Two of the world's most poisonous mushrooms belong to this genus are the *Cortinarius rubellus* (*C. rubellus*) and the *Cortinarius orellanus* (*C. orellanus*). The deadly web cap contains the toxin orellanin and orellin, while the fool's web cap has orellanin and cortinarin toxins.

C. rubellus, also known as the deadly web cap, produces brown- to orange-color spores. Morphologically, it has a rusty-brown- to orange-color pileus; gills are free and not connected to the stalk. The pileus of this mushroom looks similar to other edible mushroom species. It is highly lethal as even a small dose of it can lead to death. The *C. orellanus* is also known as fool's web cap. It has rusty-brown- to orange-color spores, concave-shaped pileus, and gills are somehow similar to the deadly web cap. It can be best grown in the acidic or alkaline soil, and hence, easily found in the forest area or around trees.

9.8.3 *GALERINA SPP.*

Galerina margina, also known as the autumn skullcap, is a well-known species under this genus. It grows throughout the world on dead and decaying wood. Morphologically, it looks similar to most of the edible species. It contains toxin, that is, amatoxin which has a very lethal effect on the human body.

9.8.4 *LEPIOTA SPP.*

The deadly dapperling (*Lepiota* spp.) mushroom is found in forests of Europe and North America. It produces lethal toxin, that is, amatoxin which can cause severe damage to liver which finally leads to death. This mushroom is responsible for the maximum proportion of all the deaths in the world caused by mushroom poisoning.

9.9 DISEASES OF MUSHROOM AND THEIR MANAGEMENT

Like a crop plant, mushrooms are also subjected to different diseases during cultivation. Both primary and secondary factors may lead to different diseases during cropping. The major causes for bacterial and fungal diseases likely to come from using infected compost and substrates used for cultivation. Most of the viral diseases are likely to get spread from spores. Mushroom cultivation is severely affected from diseases in terms of losses, and thus proper management is needed throughout the cropping period (Garcha, 1978).

9.9.1 FUNGAL DISEASES

9.9.1.1 COBWEB

Common Names: Hypomyces mildew disease and soft rot disease.

Introduction:

The cobweb disease is a major fungal-borne disease caused by the cluster of fungi and also known as mildew, soft decay, *Hypomyces* mildew, etc. The soft rot of fruit is the typical pattern of damage. It has caused widespread crop losses in the USA and Europe over the last many years. Merat (1821) showed that disease is caused by the collective action of *Botrytis dendroides* and *Cladobotryum* spp. This disease can occur during any stage of crop development and can cause severe damage (Hotson, 1917; Sharma et al., 1992).

Casual Organism: *Cladobotryum dendroides* (syn.: *Dactylium dendroides*).

Perfect stage: *Hypomyces rosellus.*

Affected Mushrooms species: Button mushroom.

Pathogen:

The hyphae formed are prostrate, septate, and branched. Conidiophores are erect, single, and elongate. They look pointed at the base and are septate having a diameter of approximately 28 × 12.5 μm. The mycelium of the pathogen is grayish white when young but turns reddish as it ages. Conidia are relatively large and multicellular but are readily dispersed in air (Goltapeh et al., 1989).

Symptoms:

The fungus grows rapidly on the casing soil and develops the cobweb-like mycelium structure on the mushroom.

At an early stage, the mushroom appears like a cotton ball, and then it converts to a soft rotting mass upon maturity. The colonized surface affected by fungus looks pale brown in color. In severe cases, the whole mushroom bed looks like yellow to red. At the final stage, affected mushrooms turn brownish black in color and rot (Upadhyaya et al., 1987).

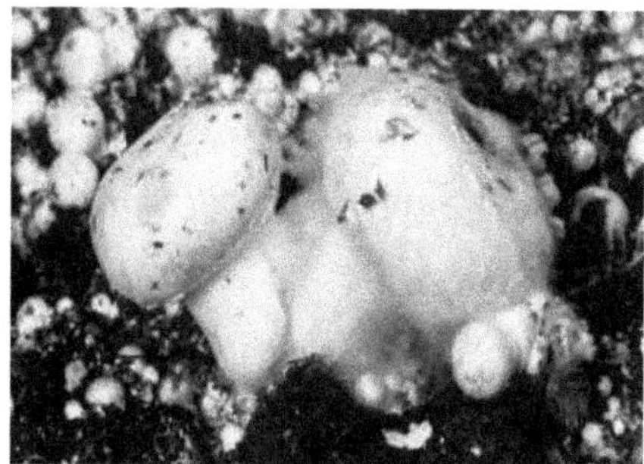

FIGURE 9.2 Cobweb disease.

Disease Cycle:

The disease is soil borne as well as air borne in nature. The wet surface with high humidity favors the establishment of the disease. The temperature around 25 °C and relative humidity at 90% are the most favorable conditions for pathogen's survival. As the fungus is inhabitant in nature, mushrooms are likely to get infected through spores and mycelium found on crop debris as well as via farm labors. Spores are air borne and may contaminate plastic containers and spread rapidly from one place to another (Seth and Dar, 1989).

Favorable Conditions:

High humidity and temperature ranging from 19 °C to 22 °C during the cropping period are the main factors that favor the growth of fungus.

Management:

1. Use only properly pasteurized casing soil and disinfected appliances (Seth and Dar, 1989).
2. Maintain strict hygienic conditions in the bed area.
3. Reduce the humidity of the farm to 80% to prevent the spreading of the disease.
4. In order to minimize air-borne spores, cover the affected mushrooms with paper.
5. For the effective control of pathogen, spray 5% formalin solution at 0.5-1.0 l/m^2 area after casing (Sharma et al., 1997).

- Spray Dithane Z-78 (25%) 3 times at 10-day intervals to control the pathogen.

6. Apply zineb (0.3%), carbendazim (0.05%), or benomyl (0.15%) immediately after casing for effective control of the disease.

9.9.1.2 DRY BUBBLES

Common Names: *Verticillium* disease, brown spot, fungus spot, and La mole.

Introduction:

It is also one of the important fungal diseases of mushrooms. It spreads very rapidly in the farm and can destroy the whole crop in no time if left uncontrolled.

Casual Organism: *Verticillium fungicola.*

Affected Mushrooms Species: Button mushroom.

Pathogen:

Verticillium species produce thin-walled sticky conidiospores that help the pathogen to survive for a longer period. Conidia produced by the pathogen are single celled, thin walled, oblong to cylindrical in shape, and verticillately branched. Conidiophores are slim and tall residing in clusters surrounded by sticky-gum-like secretion produced from the pathogen (Gularia, 1976).

Symptoms:

The pathogen develops varied symptoms according to the stage of development. The most characteristic symptom is the appearance of light-brown spots on the pileus. At the initial stage, fruiting bodies are deformed, and the pileus is tilted. At a later stage, the pileus reduces in size and turns leathery. A heavy spore mass of dark brown color is observed on the pileus. In the severe stage, downward spotting on the cap, mis-shaped caps, and poor growth of the stem are observed. Initially, the growth of pathogen of white color is observed on casing soil which later on turns gray to yellow. Sometimes, due to infection, the mushroom look like an onion (Earana, et al., 1991; Seth et al., 1973).

Disease Cycle:

The fungus is soil borne in nature and spores are likely to survive up to one year under the wet soil condition. The pathogen is introduced in the farm by infected casing soil. The primary source of infection appears to be contaminated casing soil and the disease spreads through mites and flies. In certain cases, the use of contaminated appliances and dirty hands can also cause infection.

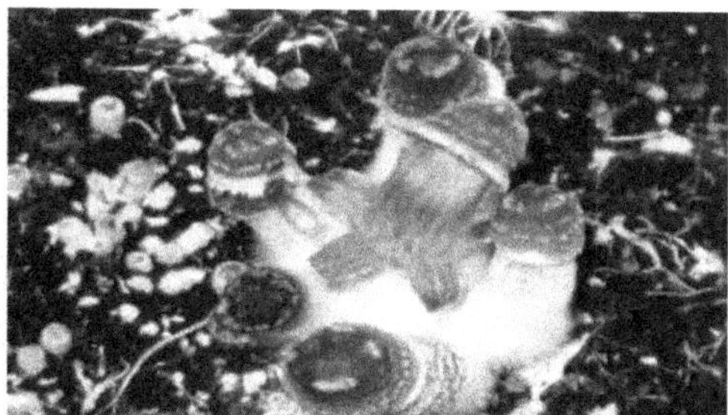

FIGURE 9.3 Dry bubble disease.

Favorable Conditions:

High humidity in the farm area and not maintaining proper hygiene during farm operations are some of the important factors favoring the pathogen. Once the infection starts, the symptoms develop within 10 days only. The maximum pathogen activity is observed at the 20 °C-24 °C range. Furthermore, lack of proper air circulation, delay in picking, and reduction in optimum temperature favor the development of the pathogen and spread it in the farm.

Management:

1. Use only sterilized casing soil.
2. Fungal infection can be avoided by maintaining proper hygiene and sanitation measures in the farm operations.
3. Controlling flies and mites helps in reducing further spread of the disease.
4. Spraying of prochloraz manganese during severe infection helps in managing the disease (Nair and Macauley, 1987).
5. Three sprays with 0.25% diathane M-45 @ 10-day intervals help in reducing infection.
6. Application of zineb (0.3%), carbendazim (0.05%), and benomyl (0.15%) immediately after casing is also advisable for managing the pathogen.

9.9.1.3 WET BUBBLES

Common Names: *Mycogone* disease, La mole, white mold, bubble.

Introduction:

It is a widely spread mushroom disease throughout the world. It was firstly reported in France in 1888 after observing destructive losses in mushroom farms. The disease later affected mushroom farms in the USA and United Kingdom (Nielson, 1932). In India, the first incidence was noticed in Jammu and Kashmir in 1978 (Kaul et al., 1978). Later, the disease prevailed in various states such as Himachal Pradesh, Maharashtra, and Haryana (Sharma, 1994a, 1994b; Bhatt and Singh, 2000a, 2000b).

Casual Organism: *Mycogone perniciosa* (*M. perniciosa*).

Perfect Stage: *Hypomyces perniciosa.*

Affected Mushrooms Species: White button mushroom and paddy straw mushroom.

Pathogen:

The pathogen produces the white-colored septate mycelium, is compact, and also has branched hyphae. Conidiophores are short and slender in appearance. The pathogen produces two spore forms: one a single-celled, thin-walled conidia having a shorter life and aleuriospore which is doubled cell with a thick wall.

Symptoms:

The disease affects the young mushroom at the pinhead stage. Due to the malformation of the stipe and pileus, it forms the monstrous-shaped body. The short and fluffy white mycelia are formed on infected mushrooms (Sharma and Kumar, 2000b). At the advanced stage, the mushroom body releases liquid secretion which produces very bad odor, and finally, the whole mushroom body becomes rotted. Due to putrefaction, the mushroom produces bad smell along with liquid exudates of brown color (Hsu et al., 1993).

Disease Cycle:

Air-borne conidia and aleuriospores disseminate further with the help of air, vectors, that is, mites/flies and through water sprinkled on the body. Aleuriospores have very long survival ability and can remain viable for 3–4 years. Conidia prefer to survive in soil for a longer period before further infection.

FIGURE 9.4 Wet bubble disease.

Favorable Conditions:

The temperature range nearly 25 °C and pH of 6.0 mostly favor growth, sporulation, and germination of conidia (Hsu and Han, 1981). Higher humidity in casing soil and improper hygienic measures favor the establishment of the pathogen.

Management:

1. Maintaining proper humidity within farm can help in managing the disease.
2. Wuest and Moore (1972) suggested that the exposure of casing soil to steam (produced above 54.4 °C) for 15 min can help in eliminating *M. perniciosa*.
3. Use of disinfected casing soil and appliances helps in mitigating the pathogen.
4. Collection and destruction of all infected mushrooms from farm help avoid further spread.
5. Spraying of benomyl @ 4 g/m^2 immediately after casing is recommended for avoiding the pathogen (Stanek and Vojtechovska, 1972; Gandy, 1974).
6. Application of zineb (0.3%), carbendazim (0.05%), and benomyl (0.15%) post casing is also advisable for managing the pathogen.

9.9.2 COMPETITOR MOLDS AND WEED FUNGI OF MUSHROOMS

Apart from major diseases, the weed and competitor fungi are another threat to mushroom cultivation. Higher temperature (>22 °C) required for the optimum growth of the mushroom also favors the competitor and weed fungi. The important competitors and weeds of mushrooms are as follows.

9.9.2.1 FALSE TRUFFLE

Common Name: Truffle disease.

Introduction:

This disease is also known as "Truffle disease." It is one of the most dangerous diseases for button mushroom species. Failing to manage the disease may reduce the mushroom yield up to 75%-80%. The pathogen was first reported in Ohio, USA (Lambert, 1930) and later described as *Pseudobalsamia microspora* (Diehl and Lambert, 1930).

Casual Organism: *Diehiliomyces microsporus.*

Affected Mushrooms Species: Button mushroom.

Pathogen:

The fungus produces oval-shaped fleshy ascocarps of white color which turns brown to red upon maturity. The pathogen is sporadic in nature, but when it occurs, it can persist and then difficult to eradicate. The pathogen competes with the mycelium for survival and causes the death of mycelia.

Symptoms:

The pathogen mycelium is initially white in color which turns to pale pink in color at a later stage. The fungus producing the truffle that is identical to the mycelium, but it is not the true mycelium. At the initial stage, it has the same color as that of the mushroom, but later on, the color changes to reddish brown. A typical acidic smell releases from the mushroom room gives an indication of the disease. The pathogen can easily spread through soil and via farm operations operating within the farm. When soil is used as casing, it is likely to prevail easily (Kumar and Sharma, 1988a, 1988b).

Disease Cycle:

The disease is primarily caused by ascospore which further disseminates by infected plant parts and also along with drainage water.

FIGURE 9.5 False truffle disease.

Favorable Conditions:

The pathogen grows at 28 °C-30 °C temperature. The temperature above the optimum range, contaminated casing, and following unhygienic practices may help in establishing the pathogen.

Management:

1. Always use disinfected and sterile casing soil for production.
2. Use a concrete floor rather than the soil surface for the preparation of the compost which helps in maintaining temperature during composting (Sharma, 1992).
3. Maintaining a hygienic condition in farm is must to manage the disease effectively.
4. The temperature during the cropping period is maintained below 18 °C to avoid the growth of the truffle as the pathogen is more active at a higher temperature range (Sharma and Sharma, 2008)..
5. Application of formaldehyde @ 2% helps in managing infection at the initial stage of infection (Sohi, 1988).

9.9.2.2 GREEN MOLD

Common Names: *Trichoderma* spot, *Trichoderma* blotch, and *Trichoderma* mildew.

Introduction:

Green color grows on a substrate usually known as the "green mold." The first incidence of this disease in a mushroom compost was recorded by

Kligman (1950). The *Trichoderma* genus has various species which work as competitors to the button mushroom, that is, *Trichoderma viride* (*T. viride*), *Trichoderma harzianum* (*T. harzianum*), *Trichoderma atroviride* (*T. atroviride*), *Trichoderma koningii* (*T. koningii*), *Trichoderma hamatum* (*T. hamatum*), *Trichoderma logibrachiatum* (*T. logibrachiatum*), and *Trichoderma pseudokoningi* (*T. pseudokoningi*). The *T. harzianum* is the most problematic among all the species. Due to this mold, the production is very much affected qualitatively as well as quantitatively.

Casual Organism: *Trichoderma* spp. *T. viride, T. koningii, T. hamatum, T. harzianum, T. atroviride, T. pseudokoningii, T. logibrachiatum. Penicillium cyclopium, Aspergillus* spp.

Affected Mushrooms Species: Button mushroom.

Pathogen:

The genus *Trichoderma* has a number of different species and strains responsible for producing the green mold. The notable species are *T. atroviride, T. aureoviride, T. hamtum, T. harzianum*, and *T. viride*. The fungus has a parasitic nature and has an ability to produce toxins. This fungus produces green to white spores which later turn to dark green.

Symptoms:

The formation of green mold on the substrate or casing layer is the prime symptom of the pathogen. In addition to this, the appearance of dark brown or green lesions on the stipe as well as on the pileus is also the indication of the fungi. In a severe condition, the infected mushroom is likely to get deformed and rotted. In some cases, the formation of brown spots on the pileus is also observed (Sharma and Vijay, 1993).

Disease Cycle:

The soil-borne fungi enter through an infected fruiting body and through a contaminated substrate and further spread with the help of mites in the mushroom farm.

Favorable Conditions:

Wet surface with high humidity, low compost, and pH value <7.0 are the key factors for pathogen establishment. Infection further develops from dead mushroom tissues.

FIGURE 9.6 Green mold.

Management:

1. Maintaining proper hygiene at different stages of cultivation reduces the chances for the occurrence of the fungus.
2. The compost material should be pasteurized and sterilized prior to utilizing for cropping helps in managing the pathogen.
3. Spraying of formalin (2%) on casing soil is an effective remedy for minimizing the pathogen activity (Sharma et al., 1999).
4. Spraying infected beds with mancozeb 0.2% or Bavistin or 0.1% Dithane Z-78 0.2% or calcium hypochlorite 15% to manage the establishment of the pathogen (Gularia and Seth, 1977).

9.9.2.3 OLIVE-GREEN MOLD

Introduction:

Chaetomium olivaceum (*C. olivaceum*) as the casual organism of the olive-green mold was first reported in India by Gupta et al. (1975) from Kasauli, Himachal Pradesh. Later, Thapa et al. (1979) reported about the olive-green mold (c.o. *Chaetomium globosum*) from Himachal Pradesh, Delhi, and Mussoorie. This mold provided yield losses up to 54% in a serious condition in button mushroom species (Sharma and Vijay, 1996).

Casual Organism: *Chaetomium* spp. (*C. olivaceum* and *C. globosum*).

Affected Mushrooms Species: Button mushroom and oyster mushroom.

Pathogen:

The pathogen produces the mycelium that is grayish white in color. The mycelium further produces thin, membranous, and scattered perithecia. Ascospores produced by the pathogen are mostly having dark brown color.

Symptoms:

The symptoms start to appear after 10 days of spawning. The formation of grayish color on mushroom beds gives an indication of the fungus. After initiation, the growth of spawn is largely affected and delay in fruiting is also observed. At a later stage, the grayish white bed turns into olive green and produces foul odor (Kumar and Sharma 1988b).

FIGURE 9.7 Olive-green mold.

Disease Cycle:

The air-borne fungus is introduced to the farm through compost and casing soil. Ascospores are further spread in the farm by air and materials used in cultivation (Sharma, 1992).

Favorable Condition:

An anaerobic condition with temperature above 30 °C favors the growth of pathogen in the mushroom farm.

Management:

1. Use of hygienic and quality compost may help in avoiding the pathogen.
2. Take extra care while preparing compost and its pasteurization as it is the most favorable factor for the pathogen to initiate infection.

3. Application of nitrogenous ingredients such as chicken manure, urea, and ammonium sulphate should not be added at a later stage of composting (Kumar and Sharma, 2000).
4. Give sufficient time for compost to get fermented.
5. Spraying of zineb (0.2%) helps in avoiding the secondary spread of the pathogen.
6. Spraying of the systemic fungicide like benomyl or carbendazim can help in minimizing the mold growth in the bed (Doshi et al., 1991).

9.9.2.4 BROWN PLASTER MOLD

Introduction:

An occasional mold appears in the farm but has ability to give yield losses up to 75% if not managed properly. The causal organism *Papulaspora byssina* of brown plaster mold was first reported on horse dung compost from Missouri (Hotson, 1917). Later, Charles and Lambert (1933) observed drastic loss in the mushroom yield due to this fungus. In India, the incidence of the brown plaster mold was first reported in the white button mushroom and causing yield loss approximately 90%-92% by Munjal and Seth (1974).

Casual Organism: *Papulospora byssina* Hots.

Affected Mushrooms Species: Button mushroom and oyster mushroom.

Pathogen:

The mycelium is septate and brown in color. The spherical bulbils produced by the pathogen attacked on host hyphae and lead to the death of the mycelium at later.

Symptoms:

The white-colored mycelial growth is found on compost and casing soil. This white color further develops and changes to brown patches. Under severe conditions, the whole bed turns to rust color. The pathogen produces brown-colored sclerotia or bulbils which suppresses the growth of the mycelium.

Disease Cycle:

The primary source for infection comes from various sources such as, casing soil, farm labors, and air-borne spores.

FIGURE 9.8 Brown plaster mold.

Favorable Conditions:

Utilization of wet and contaminated compost for cropping favors the pathogen. The pathogen is also favored by higher temperature during spawning and cropping. The gypsum plays a key role in composting and casing. The compost prepared with low gypsum likely to favor the growth of the pathogen. Even a more greasiness and wet compost in casing are also a boosting factor for the establishment of the pathogen (Dar and Seth, 1981).

Management:

1. Make compost in a proper and hygienic way to reduce chances of infection.
2. Use the optimum amount of gypsum during composting as the pathogen favors its growth at a lower level of gypsum.
3. Make sure that the compost contains an optimum amount of water, that is, not too wet.
4. Application of 2% formalin and 4% formalin is recommended by Munjal and Seth (1974) and Seth and Shandilya (1975), respectively, for effective management.
5. Application of systemic fungicides @ 0.1% concentration also helps in proper management of mold.

9.9.2.5 YELLOW MOLD/MAT DISEASE/VERTE-DE-GRIS

Introduction:

The yellow mold was first reported from mushroom caves in France by Constantin (1892). In India, the first incidence was reported from Jammu &

Kashmir (Kaul et al., 1978). The production of yellow growth in mushroom beds by a group of related fungi is also another important disease condition in mushroom cultivation providing moderate yield losses (Sohi and Upadhyaya, 1989).

Casual Organism: *Myceliophthora lutea*, *Chrysosporium luetum*, and *Chrysosporium Sulphureum*.

Affected Mushrooms Species: Button mushroom.

Pathogen:

Myceliophthora lutea produces a white color mycelium that later turns to yellow with dull white sporulation. It also produces septate and branched hyphae. Three different kinds of spores produced by pathogen are smooth spores, smooth- and thick-walled chlamydospores, and thick-walled spiny chlamydospores.

Symptoms:

The pathogen produces various symptoms such as the formation of yellow molds below the casing (Mat disease), formation of round spots in the compost (confetti), and the whole compost bed got affected by pathogen (Vert-de-girs). The appearance of yellow spots of fungi having white edges is the most typical symptom of this mold. At the initial stage, the fungus is very identical to the mushroom and cannot distinguish easily. But, at a later stage, the fungus changes its color to yellow with strong smell. Due to the yellow mold, most of mushroom pins die at a very early stage of their life (Guleria et al., 1987).

Disease Cycle:

The pathogen is introduced in farm beds via air, manure, compost, and contaminated farm materials which then further spread via water, cropping tools, and vectors, that is, mites.

Favorable Conditions:

Environmental factors such as temperature and relative humidity play a great role in favoring the establishment of the pathogen. The pathogen is most active at 20 °C temperature and at 70% relative humidity.

Management:

1. Maintaining a strict hygienic condition in the farm helps in avoiding the pathogen.

2. Regularly disinfect the farm materials and appliances with formalin (2%).
3. Spraying of systemic fungicide such as benomyl or carbendazim is advisable for managing mold establishment in beds (Seth and Bhardwaj, 1989).

9.9.2.6 SEPEDONIUM YELLOW MOLD

Introduction:

The first incidence of the pathogen in India was reported by Thapa et al. (1991a, 1991b). They also noted the yield loss from 5% to 20% by this mold. Later, Vijay et al. (1993) reported another species, that is, *Sepedonium maheshwarianum* (*S. maheshwarianum*).

Causal Organism: *Sepedonium chrysosporium* (Bull.) Fries.
S. maheshwarianum Muker. (*Hypomyces chryaosporium* TuU.)

Pathogen:

Initially, the pathogen produces white spores that turn yellowish upon maturity. The hyphen is branched, thick, and septate in nature. Conidiophores are erected and bear spores singly as well as terminally on the branches. Conidia are thin walled and ellipsoid in nature.

Symptoms:

The formation of white color initially which later turns to yellowish tan is a typical symptom of this pathogen. The symptoms normally found in the lower layers of the compost. The pathogen produces toxins that inhibit the growth of spawn from the compost (Vijay et al., 1993).

Disease Cycle:

The primary source of infection comes from casing soil and contaminated equipment. Spores are further spread by unhygienic farm operations and air. The compost with high nitrogen content are reported to favor the mold development (Vijay et al., 1993).

Management:

1. A proper method for composting, that is, hygienic way helps in eliminating the pathogen from the mushroom farm.

2. Installation of air filters during spawning helps in avoiding spores.
3. Addition of carbendazim @ 0.5% in compost with formalin @ 2% is very effective in managing the pathogen (Vijay et al., 1993).

9.9.2.7 WHITE PLASTER MOLD

Introduction:

This disease has been reported to occur commonly in different parts of India by several workers (Kaul et al., 1978; Garcha et al., 1987). This mold is mostly observed in mushroom farms where long composting methods are adopted (Bhardwaj et al., 1987). It is one of the important molds in button mushroom as it produces greater losses once established.

Casual Organism: *Scopulariopsis fimicola.*

Affected Mushrooms species: Button mushroom.

Pathogen:

The mycelium of the pathogen is septate, conidiophore short, branched, and borne irregularly as lateral branches of hyphae. Annellospores are ovate, globose, round, showing truncation, buff to avellanos in mass, and occur in chains or clusters. Conidia are short and branched. The pathogen produces ovate and round annellospores which appears in clusters or chains.

Symptoms:

The appearance of white growth on the compost or casing soil is the typical symptom of the mold. At a later stage, the white color changes to light pink and gives ammonic smell. In its sever stage, the spawn growth reduced and sometimes leads to whole crop failure too.

Favorable Conditions:

The higher pH (>8) during compost is an ideal favorable condition for the establishment of the pathogen.

Management:

1. Care should be taken while preparing the compost. The pH lower than the neutral range is must.
2. The compost should be of excellent quality and well pasteurized before use.

3. Addition of gypsum during composting helps in managing the white plaster mold.
4. Spreading of the compost may help in eliminating ammonic smell from beds.
5. Spray benomyl (0.1%) and formalin (4%) post removal of mold growth from beds.

FIGURE 9.9 White plaster mold.

9.9.2.8 LIPSTICK MOLD

Introduction:

The disease is also known as the "red lipstick disease." It was first reported in India from Punjab (Garcha et al., 1987) and HP (Sohi, 1986, 1988).

Causal Organism: *Sporendonema purpurascens*

Pathogen:

The white mycelium with frosty appearance, septate and segmented hyphae, and short cylindrical spores with red pigment are the typical characters of the pathogen.

Symptoms:

The pathogen grows as a crystal white color compost at the initial stage. Upon maturity, the color changes to pink, and at the severe stage, the color changes to cherry red or orange. The loose compost favors the establishment of the pathogen. This disease mostly comes as a secondary disease along with viral diseases of mushrooms.

Disease Cycle:

The source of primary infection comes from casing soil. The secondary spread is carried out by farm labors as well as through water.

Management:

1. Maintaining proper hygiene is must to avoid this disease. Using a pasteurized compost may help in eliminating the pathogen.

9.9.2.9 INKY CAPS

Common Names: Ink weed and wild mushroom.

Introduction:

The first incidence was reported from mushroom beds in northern India (Kaul et al., 1978). This condition is mainly introduced in the farm through a nonpasteurized compost. Due to inky caps, the overall weight of fruits as well as total yield is drastically reduced.

Casual Organism: *Coprinus* spp.

Affected Mushrooms Species: Paddy straw mushroom, button mushroom, milky mushroom, and oyster mushroom.

Pathogen:

Coprinus spp. has a cylindrical pileus which is white in color and having scales on the surface that disappears after some time. The gills are pink at first that later changes to black.

Symptoms:

The cap appears creamy white in color at the initial stage which changes to bluish black at a later stage. The cap also covers with a plenty of white-colored scales which may disappear after some time. At the advanced stage, caps are covered with black and shiny growth. The pathogen is found mostly in clusters and grows quickly under favorable conditions.

Favorable Conditions:

In most cases, the presence of high humidity and addition of more amount of nitrogenous supplements introduce fungus in mushroom beds. An inadequate amount of gypsum in manure help in the establishment of the pathogen.

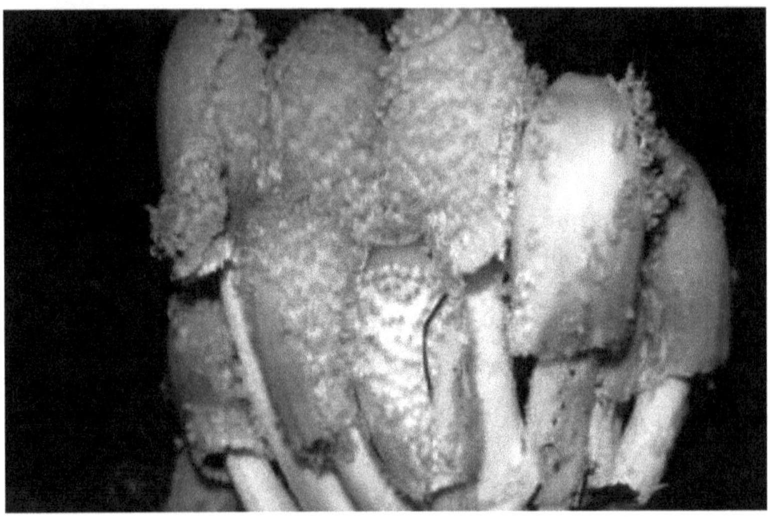

FIGURE 9.10 Ink caps.

Disease Cycle:

The fungus gets introduced in the farm through air, contaminated compost, and casing soil. Spores are further disseminated through infected bodies and appliances used in cropping.

Management:

1. Always use a pasteurized compost for cropping.
2. Maintain the humidity level in the farm as higher humidity may favor the growth of the pathogen.
3. Collect and destroy infected fruit bodies from the farm to control further spread.

9.9.2.10 OTHER COMPETITORS AND WEED MOLDS

Mushroom is also affected by various other molds that produce a little to moderate damage. Various molds such as cinnamon brown mold (c.o.: *Chromelosporium fulva*; white mycelium growth on casing soil which later turns to cinnamon brown in color), pink mold (c.o.: *Cephalothecium roseum*; white mycelium growth on casing soil which later changes to pink in color), black whisker mold (c.o.: *Doratomyces* spp.; black powdery spores appear on the compost which appeared as smoke when disturbed), black mold (c.o.: *Aspergillus* spp.; black color mold observed in casing soil.), pin mold (c.o.: *Rhizopus* spp.; tall aerial hyphae adorned with black-headed pins), lipstick mold (c.o.: *Sporendonema purpurascens*; at first, the mold appears as white crystalline specks, which later changes to pink, cherry red, or orange color mat) and Lilliputia mold (c.o.: *Lilliputia rufula*; a spawn competitor mold that appears in chicken manure is used for compost making) (Seth and Manjul, 1981; Sharma and Sharma, 2000).

These molds are favored by unhygienic practices and nonpasteurized substrates. Therefore, maintaining strict hygiene and using pasteurized substrates can easily eliminate such molds from the farm.

9.9.3 BACTERIAL DISEASES

Bacteria are another important group of microorganisms that produce a variety of diseases in mushrooms. Important bacterial diseases of different mushroom species are discussed in the following sections.

9.9.3.1 BROWN BLOTCH/BACTERIAL BLOTCH

Introduction:

It is also known by other names such as brown blotch and bacterial spot. The disease is found mostly in button mushroom and produces greater losses throughout the world. The incidence of the disease was first reported in America by Tolaas (1915). In India, it was first identified in the year 1976.

Casual Organism: *Pseudomonas tolaasii.*

Affected Mushrooms Species: Button mushroom.

Pathogen:

The bacterium is a Gram negative and soil borne in nature. It produces tolaasin, a toxin that gives blotch effect in the button mushroom.

Symptoms:

This disease is normally observed during the initial stage of mushroom growth. It is characterized by brown-color spots on pileus. Initially, the spots are of a light color that turns to dark upon maturity. The spots may also be visible on stem under a severe condition. The margin of the cap is also affected, and it shows irregular spots of yellow color during the advanced stage of pathogen growth.

FIGURE 9.11 Bacterial blotch disease.

At the final stage, the cap looks deformed, stunted, and produces bad odor. Finally, the cap lyses due to toxin secretion.

Disease Cycle:

The pathogen enters mushroom beds via casing soil and air which is then disseminated by water and cropping tools. Vectors such as mites and mushroom flies may also spread the disease in the farm.

Favorable Conditions:

High relative humidity (>80%), high temperature (>20 °C), and poor air circulation are important factors that favor the pathogen.

Management:

1. Follow cultivation in a strict hygienic condition.
2. Avoid gathering of water for longer time in beds as it may increase the humid condition.
3. Control the vectors, that is, mites and mushroom flies.
4. The disease can be effectively managed by applying streptomycin (200 ppm).

9.9.3.2 DRIPPY GILL/YELLOW BLOTCH

Introduction:

The disease is mostly found during the autumn and winter seasons. The whole mushroom farm can be infected, and total failure can be observed if this disease is left without cure (Jandaik et al., 1993).

Casual Organism: *Pseudomonas agarici.*

Affected Mushrooms Species: Button mushroom and oyster mushroom.

Pathogen:

It is a bacillus Gram-negative bacterium with motility and circular colonies. It shows positive results for the acid production test as well as those for oxidase and catalase. It shows negative response to nitrate reduction activity (Jandaik et al., 1993).

Symptoms:

The pathogen attacks gills at a very early age. Due to infection, the gills show brown-color patches from which creamy secretion releases. The caps also show variation in color due to the pathogen and turn yellow or orange as well as remain stunted. This condition is mainly caused when proportion of relative humidity exceeds 90% (Gupta et al., 1975; Mallesha and Shetty, 1988).

Disease Cycle:

The pathogen introduces through infected caps and gills which may further disseminates in farm via water, cropping tools and vectors.

Favorable Conditions:

Higher humidity and high temperature are the promising factors that favors pathogen establishment. If water remains on mushrooms body for longer period of time than the chances of this disease increases.

Management:

1. Use only healthy spawn material for cultivation.
2. Maintain proper hygiene on farm to eliminate chances of the incidence.
3. High relative humidity favors the growth of pathogen. Therefore, maintain the humidity level in the farm during the cropping period.

4. Apply chlorinated water @ 100-150 ppm of freely available chlorine (FCA) regularly to minimize losses due to the bacterial pathogen (repeated application @ 4-day intervals) (Biswas et al., 1983).
5. Apply streptomycin (0.02 %) for the management of the disease.

9.9.3.3 GINGER BLOTCH DISEASE

Introduction:

It is the latest disease found infecting cultivated mushroom species all over the world. It was first reported from the mushroom farms in United Kingdom (especially *A. bisporus*). It differs symptomatically from the blotch disease caused by *Pseudomonas tolaasii* (*P. tolaasii*). The causative organism was isolated and identified as a new member of the *Pseudomonas fluorescens* (*P. fluorescens*) complex, which can be distinguished from *P. tolaasii.*

Casual Organism: *Pseudomonas gingeri* (*P. gingeri*).

Affected Mushrooms Species: Button mushroom and oyster mushroom.

Pathogen:

P. gingeri is ascribed to the complex of *P. fluorescens*, but remains distinct from *P. tolaasii* (Nott, 1989).

Symptoms:

The infected mushroom shows small, pale yellowish-brown flecks on mushrooms which later on turns to red color. Due to the disease, the mushroom appears like ginger upon maturity. This color variation readily distinguishes *P. gingeri* infection from *P. tolaasii*. The pathogen also releases a toxin that enforces lyses of the pileus.

Favorable Conditions:

Higher humidity and higher temperature are important factors that favor the pathogen.

Management:

1. Maintenance of proper hygiene in the mushroom farm.
2. Use of a properly pasteurized compost helps in eliminating the disease.
3. Avoid high humidity and higher temperature in the farm, as both factors favor pathogen growth.

4. Spraying of streptomycin @ 0.02% helps in managing the disease effectively.

9.9.3.4 MUMMY DISEASE

Introduction:

This disease generally prevails in the United Kingdom (Wuest and Zarkower, 1991). Its occurrence in India is yet to be reported.

Casual Organism: *Pseudomonas aeruginosa.*

Affected Mushrooms Species: Button mushroom and oyster mushroom.

Symptoms:

The disease visibly appears in mushroom beds when the pin heads are developed. The pin heads appear in patches, and several of them do not grow further and remain as stuck with the casing soil. The affected fruit bodies turn grayish in color and open prematurely. In the second flush, the stalk becomes crooked and caps are tilted. The stem becomes thick and is surrounded with fluffy mycelia. Finally, the mushroom turns to tough, leathery dry and brown, appearing as a mummy.

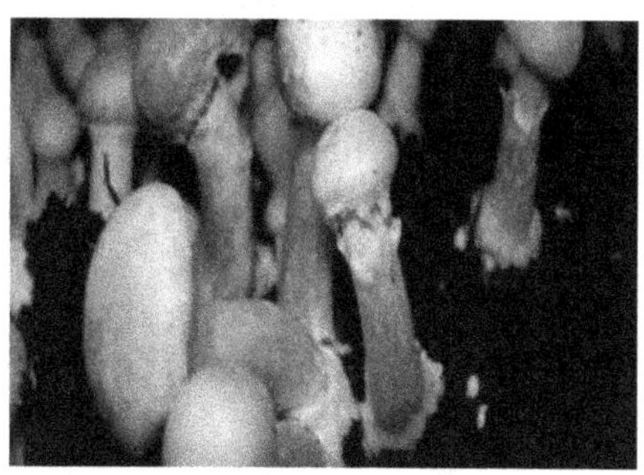
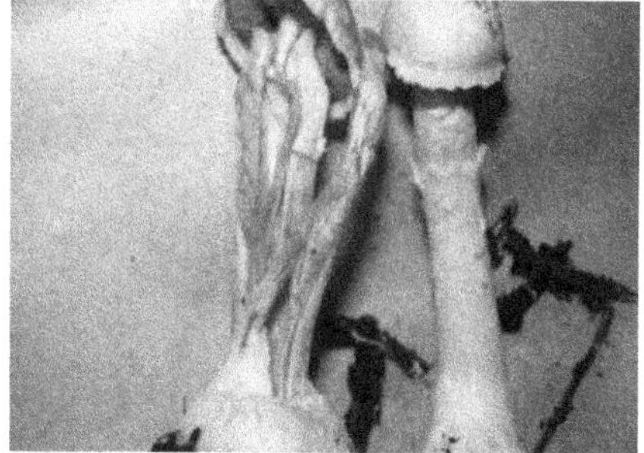

FIGURE 9.12 Mummy disease.

Management:

1. Maintaining a strict hygienic condition and sanitization help in avoiding the pathogen.
2. Collect and destroy infected mushroom bodies of the farm.
3. Treatment with formalin (0.5%) is found effective against the disease.

4. Use of properly pasteurized compost, avoidance of excessive humidity, and high temperature as well as poor ventilation in growing rooms are effective measures to prevent the disease in the farm.

9.9.4 VIRAL DISEASES

9.9.4.1 LA FRANCE DISEASE

Introduction:

La France is a term that denotes a group of viruses that are responsible to cause diseases in different mushroom species (Nair, 1973). The viruses such as MV-1 (spherical particles with diameter 25 nm), MV-2 (diameter 29 nm), MV-3 (bacilliform particles), MV-4 (diameter 35 nm), MV-5 (diameter 50 nm), club-, and rod-shaped viruses are some of the important viruses responsible for mushroom diseases (Sharma, 1995). However, in India, only two viruses, MV-2 and MV-4, have been identified and reported to cause diseases (Goltapeh and Kapoor, 1990).

Symptoms:

It is also known by other names such as "La France disease," "brown disease," "watery stipe disease," "X-disease," and "dieback disease" (Tewari and Singh, 1984). Mushrooms mature early, and then mostly grow in clusters. Mycelia that are affected by the pathogen show abnormal fruit size, shape, and structure. Pin heads have shorter size than mycelia. Sporophores appear late with less number. The stem becomes spongy and quickly turns watery. A typical smell is released from the mushroom affected by viruses (Nair, 1976).

Management:

1. Collect and destroy infected fruit bodies from the cultivation room (Raychaudhary, 1978).
2. Maintain strict hygiene in the farm (Tewari and Singh, 1985).
3. Use only disease-free seedlings as well as tolerant or resistant strains for cultivation (Sharma, 1994b).
4. Picking should be done before opening of mushrooms (Sharma and Kumar, 2000a).
5. All wooden parts of growing units must be clean and sterilized prior to use (Sharma, 1991).

6. Always clean the room and apply steam treatment before new cultivation starts (Raychaudhary, 1978).

KEYWORDS

- **mushrooms**
- **cropping**
- **morphology**
- **medicinal benefits**

REFERENCES

Bhardwaj, S. C., Jandaik, C. L. and Beig, G. M. (1987). Gliocladium virensa new pathogen of *Pleurotus* spp. *Mush. J. Tropics* 50: 97–100.

Bhatt, N. and Singh, R. P. (2000a). Chemical and biological management of major fungal pathogens *Agaricus hisporus* (Langc) Imbach. *Mush. Sci.* 15: 587–593.

Bhatt, N. and Singh, R. P. (2000b). Incidence and losses in yield by fungal pathogens encountered from the beds of *A. bisporus*. *Indian. J. Mush.* 18: 46–49.

Bhavani Devi and Nair, M. C. (1986). Popularization of *Volvariella* spp. in the tropics. In: *Beneficial Fungi and Their Utilization* (Nair M. C. and Balakrihnan, eds.). Jodhpur: Scientific Publishers, pp. 25–33.

Biswas, P., Sasrkar, B. B., Chakravarty, D. K. and Mukherjee, N. (1983). A new report on bacterial rotting of *Pleurotus sajor-caju*. *Indian Phytopath.* 36: 564.

Biswas, S., Datta, M. and Nagachan, S. V. (2012). Mushrooms—a manual for cultivation. New Delhi: Raj Press, pp. 100–126. ISBN: 978–81-203–4495-5.

Bonnefons, N. de (1650). *Le Jardinier Français*, Paris.

Callow, E. (1831). Observations on methods now in use for the artificial growth of mushrooms, with a full explanation of an improved mode of culture, Fellowes, London. (Reprinted, 1965, by W. S. Maney and Son Ltd: Leeds.).

Charles, V. K. and Lambert, B. (1933). Plaster mould occurring in mushroom green houses. *J. Agric. Res.* 12: 1089–1093.

Constantin, J. (1892). Recherches sur la môle, maladie du champignon de couche. *Revue Générale Botanique* 4: 401–406.

Dar, G. M. and Seth, P. K. (1981). Studies on brown plaster mould and its control. *Indian J. Mush.* 7: 60–63.

Diehl, W. W., Lambert, E. B. (1930). A new truffle in beds of cultivated mushrooms. *Mycologia* 22, 223–226.

Doshi, A., Sharma S. S. and Trivedi, A. (1991). Problems of competitor moulds and insect-pests and their control in the beds of *Calocybe indica* P&C. *Adv. Mush. Sci.* 2: 57.

Earana, N., Mallesha, B. C. and Shetty, K. S. (1991). Brown spot disease of oyster mushroom and its control. *Mush. Sci.* 11(2): 293–311.

Gandy, D. G. (1974). Weed moulds. *Mush. J.* 23: 428–429.

Garcha, H. S. (1978). Diseases of mushroom and their control. *Indian Mush. Sci.* 1: 185–191.

Garcha, H. S. (1984). *A Manual of Mushroom Growing*. Ludhiana: PAU Publication, p. 71.

Garcha, H. S., Khanna, P. K. and Sandhu, G. S. (1987). Status of pests in the cultivated mushrooms in India. In: *Cultivating Edible Fungi* (Wuest, P. J., Gela Royse, D. J. and Beelemen, R. B.). The Netherlands: Elsevier Science Publisher, pp. 649–665.

Goltapeh, E. M., Jandaik, C. L., Kapoor, J. N. and Prakash, V. (1989). *Cladobotryum verticillatum*—a 77 new pathogen of Jew's ear mushroom causing cobweb disease. *Indian Phytopath.* 42: 305.

Goltapeh, E. M. and Kapoor, J. N. (1990). VLP's in white button mushroom in North India, *Indian Phytopath.* 43: 254 (Abstract).

Guleria, D. S. (1976). A note on the occurrence of brown blotch of cultivated mushrooms in India. *Indian J. Mush.* 2(1): 25.

Guleria, D. S. and Seth, P. K. (1977). Laboratory evaluation of some chemicals against *Cephalothecium* mould infecting mushroom beds. *Indian J. Mush.* 3(1): 24–25.

Guleria, D. S., Thapa, C. D. and Jandaik, C. L. (1987). Occurrence of diseases and competitors during cultivation of *A. bitorquis* and their control. *Natl. Symp. Adv. Mycol.* 1: 55–56.

Gupta, G. K., Bajaj, B. S. and Suryanarayana, D. (1975). Studies on the cultivation of paddy straw mushroom (*Volvariella volvacea* and *V. diplasia*). *Indian Phytopath.* 23: 615–620.

Hotson, J. W. (1917). Notes on bulbiferous fungi with a key to described species. *Bot. Gaz.* 64: 265–284.

Hsu, H. K. and Han, Y. S. (1981). Physiological and ecological properties and chemical control of *Mycogone per-niciosa* Magnus causing wet bubble in cultivated mushrooms. Mushroom Science 11: 403–425.

Jandaik, C. L., Sharma, V. P. and Raina, R. (1993). Yellow blotch of *Pleurotus sajor-caju* (fr) Singer—a bacterial disease new to India. *Mush. Res.* 2: 45–48.

Kannaiyan, S. and Prasad, N. N. (1978). Production of paddy straw mushroom in India—review. *Indian Mush. Sci.* 1: 287–293.

Kaul, T. N., Kachroo, J. L. and Ahmed, N. (1978). Diseases and competitors of mushroom farms in Kashmir Valley. *Indian Mush. Sci.* 1: 193–203.

Kligman, A. M. (1950). A Basidiomycete probably causing onychomycosis. *J Investigative Dermatol.* 14(1): 69–70.

Krishna, Reddy, M., Pandey, M. and Tewari, R. P. (1993). Immunodiagnosis of mycovirus infecting oyster mushroom (*P. florida*). Golden Jubliee Symp. Hort. Res. Shanging Scenario. May 24–28, Bangalore, p. 248.

Kumar, S. and Sharma, S. R. (1998a). New and noteworthy pests and diseases of shiitake, Lentinula edodes in India. *Indian J. Mush.* 15(1): 52–56.

Kumar, S. and Sharma, S. R. (1998b). Transmission of parasitic and competitor moulds of button mushroom through flies. *Mush. Res.* 7(1): 25–28.

Kumar, S. and Sharma, S. R. (2002). Prospects of IPM in the management of mushroom pests. In Current Vistas in Mushroom Biology and Production (Upadhyay, R. C., Singh, S. K. and Rai, R. D., eds.).Solan: MSI, NRCM, pp. 213–224.

Lambert, E. B. (1930). Studies on the relation of temperature on the growth, parasitism, thermal death points and control of *Mycogone perniciosa*. *Phytopathology.* 20: 75–83.

Mallesha, B. C. and Shetty, K. S. (1988). A new brown spot disease of oyster mushroom caused by *Pseudomonas stutzeri*. *Curr. Sci.* 57: 1190–1192.

Mérat, F. V. (1821). Nouvelle flore des environs de Paris. 2 ed. 1. *Cryptogamie*. 292 pp.
Munjal, R. L. and Seth, P. K. (1974). Brown plaster mould—a new disease in white button mushroom. *Indian Hort.* 18(4): 13.
Nair, N. G. (1973). Heat therapy of virus infected diseases of the cultivated mushroom, *A. bisporus* in Australia. *Austr. J. Agric. Res.* 24: 533–541.
Nair, N. G. (1976). Diagnosis of mushroom virus diseases. *Austr. Mush. Growers Assoc. J.* 2(5): 22–24.
Nair, N. G. and Macauley, B. J. (1987). Dry bubble disease of *A. bisporus* and *A. bitorquis* and its control by prochloraz manganese complex NZJ. *Agric. Res.* 30(1): 107–116.
Nielsen, O. (1932). Mushroom diseases. Reprinted from Gartnertidende, 2 p.
Nott, H. M. (1989). A study of ginger blotch disease of mushroom, *M.Sc. Thesis*, University of Cantebury, Christchurch, New Zealand, p. 82.
Raychaudhury, S. P. (1978). Virus diseases of mushroom, *Indian Mush. Sci.* 1: 205–214.
Seth, P. K. (1977). Pathogens and competitor of *A. bisporus* and their control. *Indian J. Mush.* 3: 31–40.
Seth, P. K. and Bhardwaj, S. C. (1989). Studies on vert-de-gris caused by *Myceliophthora lutea Coast* on *A. bisporus* and its control. *Mush. Sci.* 12(2): 725–733.
Seth, P. K. and Dar, G. M. (1989). Studies on *Cladobotryum dendroides* causing cobweb disease of *A. bisporus* and its control. *Mush. Sci.* 12(2): 71–723.
Seth, P. K., Kumar, S. and Shandilya, T. R. (1973). Combating dry bubble of mushrooms. *Indian Hort.* 18(2): 17–18.
Seth, P. K. and Munjal, R. L. (1981). Studies on *Lilliputia rufala* (Berk. and Br.) *Hauges* and its control. *Mush. Sci.* 11(2): 427–441.
Seth, P. K. and Sandaliya, T. R. (1975). Effect on different quantities of super phosphate on the productivity of *Agaricus bisporus*. *Indian J. Mushrooms*. 1(2):10–12.
Sharma, S. R. (1991). Viruses in mushrooms—a review. *Adv. Mush.* Sci. 1: 61.
Sharma, S. R. (1992). Compost and casing mycoflora from mushroom farms on northern India. *Mush. Res.* 1: 119–121.
Sharma, S. R. (1994a). Survey for diseases in cultivated mushrooms. *Annu. Rep. NRCM*. 23.
Sharma, S. R. (1994b). Viruses in mushrooms. In: *Mushroom Biotechnology* (Nair, M. C., Gokulapalan, C. and Das, L., eds.). New Delhi: Indus Publishing Company, pp. 658–685.
Sharma, S. R. (1995). Management of mushroom diseases. In: *Advances in Horticulture*, Vol. 13. New Delhi: Malhotra Publishing House. pp. 195–238.
Sharma, S. R. and Kumar, S. (2000a). Viral diseases of mushrooms. In: *Diseases of Horticultural Crops Vegetables, Ornamentals and Mushrooms* (Verma, L. R. and Sharma, R. C., eds.). Jodhpur: Scientific Publishers, pp. 166–178.
Sharma, S. R. and Kumar, S. (2000b). Studies on wet bubble disease of white button mushroom, *A. bisporus* caused by *M. perniciosa*. *Mush. Sci.* 15(2): 569–575.
Sharma, S. R., Kumar, S. and Sharma, V. P. (2008). Diseases and competitor moulds of mushrooms and their management. *Tech. Bull.* 78. https://www.semanticscholar.org/paper/Diseases-and-Competitor-Moulds-of-Mushrooms-and-Sharma-Kumar/fa466a94eba441cd425a73d39b1ad4f64337b6f6.
Sharma, S. R. and Vijay, B. (1993). Competitor moulds—a serious threat to *A. bisporus* cultivation in India. *Proc. Golden Jubilee Symp. Hort. Soc. India, Bangalore*, pp. 312–313.
Sharma, S. R. and Vijay, B. (1996). Prevalence and interaction of competitor and parasitic moulds in *A. bisporus*. *Mush. Res.* 5(1): 13–18.
Sharma, V. P. and Sharma, S. R. (2000). Mycoflora associated with chicken manure and post mushroom substrate. *Indian J. Mush.* 18: 53–56.

Sharma, V. P., Sharma, S. R. and Jandaik, C. L. (1997). Efficacy of formaldehyde against some common competitors and mycoparasites of mushrooms. *Mush. Res.* 6: 83–87.

Sharma, V. P., Sharma, S. R. and Kumar, S. (1999). Adverse effect of formaldehyde on some edible fungi and their mycoparasites. *Mush. Res.* 8(2): 23–26.

Sharma, V. P., Suman, B. C. and Guleria, D. S. (1992). *Cladobotryum verticillium*—a new pathogen of *A. bitorquis. Indian J. Mycol. Pl. Path.* 22(1): 62–65.

Sohi, H. S. (1986). Diseases and competitor moulds associated with mushroom culture and their control. *Extension Bull.* 2: 12.

Sohi, H. S. (1988). Diseases of white button mushroom (*A. bisporus*) in India and their control. *Indian J. Mycol. Pl. Path.* 18: 1–18.

Sohi, H. S. and Upadhyaya, R. C. (1989). New and noteworthy diseases problems of edible mushrooms in India. *Mush. Sci.* 12(2): 611–614.

Tewari, R. P. and Singh, S. J. (1984). Mushroom virus disease in India. *Mushroom J.* 142: 354–355.

Tewari, R. P. and Singh, S. J. (1985). Studies on virus diseases of white button mushroom in India. *Indian J. Virol.* 1: 35–41.

Thapa, C. D., Bhardwaj, S. C. and Sharma, V. P. (1991a). Occurrence of *Sepedonium* yellow mould (*Sepedoniu chrysopermum*) in mushroom (*Agaricus bisporus*) beds. In: *Indian Mushrooms* (Nair, M. C., ed.). Vellanikkara: Kerala Agricultural University, pp. 61–62.

Thapa, C. D., Seth, P. K. and Pal, J. (1979). Occurrence of olive green mould in mushroom beds and its control. *Indian J. Mushrooms*, 5: 9–13.

Thapa, C. D., Sharma, V. P. and Bhardwaj, S. C. (1991b). Occurrence of Sepedonium yellow mould in mushroom (*A. bisporus*) beds. *Adv. Mush. Sci.* 3: 22–23.

Tolaas, A. G. (1915). A bacterial disease of cultivated mushrooms. *Phytopathology* 5: 51–54.

Tourneforte, J. de. (1707). Observations sur la naissance et sur la culture deschampignons. Memoires. *Academie Royale des Science*, 1707: 58–66.

Upadhyaya, R. C., Sohi, H. S. and Vijay, B. (1987). *Cladobotryum apiculatum*—a new mycoparasite of Pleurotus beds. *Indian Phytopath* 40: 294.

Vijay, B., Gupta, Y. and Sharma, S. R. (1993). *Sepedonium maheshwarianum*—a new competitor of *A. bisporus. Indian J. Mycol. Pl. Path.* 23: 121.

Wuest, P. J. and Moore, R. K. (1972). Additional data on the thermal sensitivity of selected fungi associated with *Agaricus bisporus*. *Phytopathology*. 62(12): 1470–1472.

Wuest, P. J. and Zarkower, P. A. (1991). Mummy disease of button mushrooms. Causation, crop loss, mycosphere implications. In: *Mushrrom Biology and Biotechnology*. Chambaghat, Solan: Mushroom Society of India.

CHAPTER 10

MUSHROOM DISEASES AND DISORDERS: DIAGNOSIS AND MANAGEMENT

SACHIN GUPTA[1*], BABY SUMMUNA[2], SUDHEER KUMAR ANNEPU[3], MONI GUPTA[4], and UPMA DUTTA[5]

[1]*Division of Plant Pathology, Sher-E-Kashmir University of Agricultural Sciences and Technology of Jammu, Jammu, India*

[2]*Division of Plant Pathology, Sher-E-Kashmir University of Agricultural Sciences and Technology of Kashmir, Kashmir, India*

[3]*ICAR-Directorate of Mushroom Research, Solan, Himachal Pradesh, India*

[4]*Division of Biochemistry, Sher-E-Kashmir University of Agricultural Sciences and Technology of Jammu, Jammu, India*

[5]*Division of Microbiology, Sher-E-Kashmir University of Agricultural Sciences and Technology of Jammu, Jammu, India*

[*]*Corresponding author. E-mail: sachinmoni@gmail.com.*

ABSTRACT

Mushroom cultivation is an attentively controlled biological system; however, contamination with various biological agents is inevitable. An array of pests, diseases, and competitor molds remains a serious issue in mushroom cultivation. The management of these adverse agents to the tolerable level is a key factor to become successful in mushroom production. In an ecosystem, a spectrum of living organisms coexists and if any living component affects another living organism, they are called as biotic factors.

Biotic factors include bacteria, fungus, virus, insects, mites, nematodes, and other living organisms.

10.1 INTRODUCTION

Mushroom cultivation is an attentively controlled biological system; however, contamination with various biological agents is inevitable. An array of pests, diseases, and competitor molds remains a serious issue in mushroom cultivation. The management of these adverse agents to the tolerable level is a key factor to become successful in mushroom production. In an ecosystem, a spectrum of living organisms coexists and if any living component affects another living organism, they are called as biotic factors. Biotic factors include bacteria, fungus, virus, insects, mites, nematodes, and other living organisms. These organisms compete for nutrition, space, moisture, light, etc. with mushrooms as both shares the same habitat. As a result, efficient organism lives whereas nonefficient individuals are removed from the particular environment with the progress of time. However, the nonliving agents influencing the development as well as the growth of living beings are considered as abiotic factors. They include temperature, relative humidity, CO_2 and O_2 concentration, presence of chemicals, growing medium, nutrients, and moisture besides many other factors. Unfavorable environmental conditions adversely affect the growth and development of mushrooms. It causes abnormalities and malformation in mushrooms, thus yielding poor quality coupled with reduced yields. Hence, understanding the physiology and biological behavior of disease-causing organisms is very essential to implement the crop protection measures effectively. The reason behind the occurrence of many disease and pest problems in mushroom cultivation may be summarized as:

1. For initiation of fruiting, mushrooms require high humidity coupled with warm temperatures inside the cropping room and the same conditions are highly favorable for breeding of many pathogens and pests.
2. Being a macro fungal organism, there is a limitation on the usage of chemicals for pest and disease management in crop.
3. During compost preparation, disease-causing pathogens and pests gets readily attracted inside and/or outside mushroom houses.
4. Poor hygiene and sanitation under seasonal growing conditions.

10.2 FUNGAL DISEASES AND COMPETITOR MOLDS

10.2.1 WET BUBBLE

Introduction/Economic Importance:

Wet bubble is the most common fungal pathogen causing severe losses during white button mushroom cultivation. Besides India, it has been reported to assume serious proportions in other major mushroom growing countries such as the Netherlands, UK, China, and Taiwan. In India, Kaul et al. (1978) reported this disease for the first time from some mushroom farms in Jammu and Kashmir; however, the occurrence of this disease was later on reported from the states of Himachal Pradesh, Haryana, and Maharashtra (Sharma and Kumar, 2000; Bhatt and Singh, 2000; Sharma and Singh, 2003).

Pathogenic Fungi, Competitor Molds, and Bacterial Diseases in White Button Mushroom

Biotic Factor	Common Name of Disease/Competitor mold	Causal Organism	Source of Contamination
Fungal	Wet bubble	*Mycogone perniciosa*	Casing soil
	Dry bubble	*Verticillium fungicola*	Casing soil
	Cobweb	*Cladobotryum dendroides*	Casing soil
	Green mold	*Trichoderma* spp.	Compost and casing
	False truffle	*Diehliomyces microspores*	Compost and casing
	Brown plaster mold	*Papulospora byssina*	Compost and casing
	Yellow mold	*Myceliopthora lutea, Chrysosporium* spp.	Casing soil
	White plaster mold	*Scopulariopsis fumicola*	Compost and casing
	Inky caps	*Coprinus* spp.	
	Lipstick mold	*Sporendonema purpurascens*	Compost and casing
	Black mold	*Doratomyces microspores*	Compost
	Cinnamon brown mold	*Chromelosporium fulva*	Casing soil
Bacterial	Bacterial blotch	*Pseudomonas tolaasii*	Casing soil and airborne dust
	Ginger blotch	*P. gingeri*	Casing soil and airborne dust
	Drippy gill	*P. agarici*	Casing soil and airborne dust
	Mummy	*P. aeruginosa*	Casing soil and airborne dust

Casual Organism: *Mycogone perniciosa* (Magnus) Delacroix, Etiology

M. perniciosa produces two types of conidia: small, single celled, thin-walled phialoconidia, and much larger bicellular conidia (aleuriospores), which consist of a dark, spherical thick-walled, verrucose apical cell situated on a thin-walled basal cell. Phialo conidia are formed at the tip of phialides, whereas aleuriospores develop on short lateral hyphae. Conidiophores are branched and cylindrical (Potocnik, 2006; Glamoclija et al., 2008). *M. perniciosa* primarily spreads through casing soil and the conditions like relative humidity and temperature favor its further multiplication. The infection can be both air or water borne or may be mechanically carried by flies and mites (Garcha, 1978). Fletcher et al. (1995) showed that pathogenic isolates of *M. perniciosa* are grown slowly on agar-producing pigmented mycelium.

Symptoms:

The pathogen may infect white button mushroom at any of its developmental stage. Infection at pinhead stage results in large, undifferentiated, and irregular forms of mushroom tissue (sclerodermoid mushrooms) covered with the white and fluffy mycelium of the pathogen. Later infections lead to cap spotting and deformation of the carpophores (Fletcher and Ganney, 1968; Hsu and Han, 1981). The presence of amber liquid droplets on the surface of distorted mushrooms is also a characteristic symptom of the disease. Early development of wet bubble is usually associated with the use of infected spawn or unsterilized casing soil (Fletcher et al., 1989; Sharma and Kumar, 2005). Under dry conditions, the distorted masses remain dry in appearance when mature mushrooms get infected by this disease, base of their stalk may get affected and the cap develops symptom on parts of the gills.

Management:

1. As the pathogen causes serious losses to the white button mushroom, strict hygienic conditions in and around the cropping rooms should be maintained. The compost prepared under short or indoor method needs to be properly pasteurized at 59 °C for 6 h with sufficient aeration. However, long method of composting should be undertaken on a cement floor.
2. Pasteurization of casing at 65 °C with 65% moisture may further prevent the entry of *M. perniciosa* in the cropping rooms. During

harvesting of mushrooms, new flushes should be harvested first followed by successive flushes. Spent mushroom substrate should be properly disposed away from the cropping rooms and should be given chemical treatment or steam sterilization before disposal.
3. As soon as the fruit bodies infected by wet bubble are observed, plastic cups may be used to cover the sporophores of infected mushrooms followed by removal and proper disposal to further prevent further spread of the disease. Spray of formalin on casing surface can also be effective. However, precaution should be taken at a later stage of crop development.

10.2.2 DRY BUBBLE

Introduction/Economic Importance:

Dry bubble is a common disease encountered by commercial mushroom production units and may cause crop losses up to 20% or even more (Grogan et al., 2009). If timely action is not taken, the disease may wipe out an entire crop within 2–3 weeks. As the spores of the fungus can rest in debris and reinfect upcoming crop, the disease may cause losses for years (Berendsen et al., 2010). A poor understanding of the disease spread within and between mushroom crops has been a major contributing factor in the persistence of dry bubble problems.

Casual Organism: *Verticillium fungicola*

Symptoms:

Malthouse in 1901 reported dry bubble disease of white button mushroom for the first time. He found *Verticillium* spp. associated with this disease and observed two types of symptoms. Initially, fungal growth appeared on the casing soil, which later on spread and turned grayish yellow. Later on, light brown superficial spots are observed on the mushroom caps, which finally coalesce to become large brown blotches. This disease gets transmitted by contaminated compost and casing soil (Kumar et al., 2014), the presence of human beings, and splashes of water (Cross and Jacobs, 1969). *V. fungicola* infected mushrooms show typical thickening of stem, resulting in onion shaped fruiting bodies. Small undifferentiated masses of tissue are observed if mushrooms are infected by this fungus at an early stage of crop growth. Fruiting bodies become deformed and caps are partially differentiated. If infection occurs at a later stage, the stipes become distorted and has tilted caps. Infected mushrooms show the presence of gray white mycelium and become discolored and dry. They show small pimple like outgrowth or brown gray spots on the surface (1–2 cm) surrounded by yellow or bluish gray halo around them.

Management:

1. Maintenance of hygiene and prevention of infected inoculum to new mushroom farms forms the back bone of strategies for management and control of dry bubble disease (Berendsen et al., 2010). All the debris should be immediately removed from the farm to maintain strict hygiene. Mites and flies should be eliminated as they further spread the pathogen by transporting spores from infected to healthy mushrooms. After spawning, dichlorvas @ 30 mL/100 L water/100 m^3 bed may be sprayed to check the flies. Casing mixture should be prepared in a clean room and stored away as to prevent the introduction of the pathogen to the ensuing crop (Tsarev, 2014).
2. Spores of dry bubble may overwinter for 7–8 months under dry conditions, resulting in a reservoir of inoculum which may be infective (Berendsen et al., 2010). High temperatures above 40 °C, anaerobic conditions and low PH and may prevent the spores as these are the unfavorable conditions (Berendsen et al., 2011). Steam treatments of the cropping rooms during the time between two crops can effectively eliminate inoculum; the technique can be highly effective if the cropping rooms are completely sealed and heated (Berendsen et al., 2011).
3. Although, use of selective fungicides can be effective by reducing the amount of inoculum, but there are many concerns with usage of chemicals including sensitivity of mushroom hosts to fungicides and development of resistance to many fungicides (Berendsen et al., 2010; Marlowe and Romaine, 1982). Sporgon (Prochlorax-Manganese) is generally recommended for control of dry bubble disease. Owing to decreased sensitivity of *Lecanicillium fungicola* to sporgon, increasing concentrations of sporgon may be needed to control dry bubble disease (Marlowe and Romaine, 1982). One spray of sporogon (0.15%) after 1 week of casing is recommended.
4. Carbendazim has been found to be the most effective fungicide in reducing the disease incidence under bed conditions (Kumar et al.,

2014). Bhatt and Singh (2002) reported sporogon (0.075%) to be effective against *V. fungicola*. Efficacy of carbendazim against *V. fungicola* has been reported by Navarro et al. (2011) and Sharma and Satish (2012).

5. Sylan A15 is the reported commercial and susceptible *A. bisporus* strain to *L. fungicola* and the main partial resistance strain is MES01497. The response of partially resistant strains is similar to the hypersensitive response observed in plants where cells in the infected tissue die and encapsulate the infection (Grogan, 2012; Berendsen et al., 2013a, 2013b). Lack of pathogen recognition implies that mushrooms rely on constitutive and wound-induced defenses (Grogan, 2012; Berendsen et al., 2013a, 2013b).
6. Use of volatile 1-octen-3-ol is reported to inhibit the germination of *L. fungicola* and its increased levels can effectively control the pathogen (Berendsen et al., 2013a, 2013b). Its application also stimulates the growth of bacterial populations in the casing layer, which play major role in the onset of fruit body formation in button mushroom.

10.2.3 COBWEB

Introduction/Economic Importance:

This disease causes extensive damage by either causing soft rot or decay of fruit bodies of white button mushroom. The disease was considered to be the most severe causing huge losses to mushroom cultivation in the UK and Ireland during the 1990s (Gaze, 1995, 1996). In India, cobweb was recorded for the first time at Chail and Shimla in Himachal Pradesh (Seth, 1977) and later in Solan and Kasauli with a natural incidence ranging from 8.17% to 25.63% in 1986 (Seth and Dar, 1989).

Casual Organism:

Several species of *Cladobotryum*, including *C. dendroides*, *C. mycophilum*, *C. varium*, *C. multiseptatum*, and *C. verticillatum*, are known to be the causal agents of "cobweb disease" in white button mushroom. The genus *Cladobotryum* syn. *Dactylium* introduced into mycological literature by Nees (1817) was identified to be the imperfect stage of *Hypomyces rosellus* (Fletcher and Atkinson, 1977).

Symptoms:

At the onset of disease, a fluffy white mycelium grows over the casing or fruit body, which quickly decays and become rotten. Owing to profuse sporulation, the feathery mycelium becomes denser and occasionally acquires yellow and pink hues. Mushroom caps may also get brown or gray spotting. *C. dendroides* causes quantitative as well as qualitative losses to the white button mushrooms. Quantitative losses are caused by the characteristic gray, coarse, and cobweb like mycelial growth on the casing layer thus reducing the cropping area and wet rot of mushrooms, while qualitative losses are due to cap spotting.

Disease Cycle:

It is normally introduced into the cropping room through contaminated casing soil or aerial spores. Secondary spread of the disease occurs by air movement, pickers, water splashes, etc. Release and dissemination of conidia within mushroom cropping house facilitates the establishment of further colonies on the casing and allows spotting to develop on mushroom caps rendering them unsalable. Spotting on mushroom caps signifies the initial stages of infection, which if allowed to develop, eventually engulf the entire cap, ultimately leading to further sporulation. It is therefore essential that spotted mushrooms are identified and discarded at picking. It has been observed that high relative humidity and temperature favor the disease spread.

Management:

1. Proper sterilization of casing mixture at 60 °C for 4 h is reported to effectively eliminate the pathogen. Regular cleaning, proper removal of cut mushroom ends, and regulating temperature and humidity helps in control of disease (Sharma, 1994).
2. Single application of sporogon (prochloraz manganese) at 1.5 g a.i./m^2 of bed after 9 days of casing satisfactory controls the disease.
3. Seth and Dar (1989) obtained maximum disease control by application of Bavistin + TMTD at 0.9 and 0.6 g/m^2 followed by TBZ and Benlate (0.9 g/m^2).
4. Effective control of *C. verticillatum* was obtained by application of 0.05% carbendazim at the time of spawning followed by 0.25% mancozeb during casing followed by repeated spray of carbendazim after a fortnight (Sharma et al., 1992).

10.2.4 GREEN MOLD

Introduction/Economic Importance:

Green mold disease or weed mold is referred to as infections in mushroom cultivation due to members of the genus *Trichoderma.* Different species of *Trichoderma* have been known to limit the production of white button mushroom (Sinden and Hauser, 1953). Sinden (1971) considered the association of genus *Trichoderma* in compost as an indicator of its poor quality. Earlier green mold was considered as a minor problem associated mainly with the low quality of the compost or poor hygiene. The devastating nature of green mold could be effectively managed by modifying the composting process and improving the sanitation (Geels et al., 1988).

First green mold epidemics in the British Isles during 1985–1986 and in the late 1990s caused huge losses (Seaby, 1998). Serious losses were also reported from the Netherlands (Geels, 1997). First observations of *Trichoderma* strains in France were made in La Rioja during the winter of 1996–1997 (Garcia-Morras and Olivan, 1999; Hermosa et al., 1999).

Casual Organism: *Trichoderma viride, T. hamatum, T.harzianum, T. koingii*

Symptoms:

Trichoderma spp. has been reported to be associated with spawn, compost, and colonized grains after spawning and casing soil. Symptoms of green mold are characterized by large dense green areas on the compost and casing surface, which ultimately results in a dramatic reduction in mushroom yield (Anderson et al., 2000). Similar symptoms are seen on dead buds of mushrooms and cut stumps. Upper side of mushroom caps may turn brown. Compost, which is generally deficient in nitrogen and rich in carbohydrates, generally favors the appearance of green molds. Development of green mold in mushrooms is favored by high humidity coupled with low pH of casing mixture. Green molds may colonize the substrate or grow on the surface of the newly emerging fruit bodies of mushrooms. In the earlier stages, *Trichoderma* spp. produce whitish mycelium that is indistinguishable from that of the mushroom mycelium during spawn run, therefore it is difficult to recognize the infection at this stage (Largeteau et al., 2002). Subsequently, large patches of compost turn green rapidly as spore production of *Trichoderma* begins, which had run through the compost along with the mycelium of *Agaricus* (Seaby, 1996; Rinker, 1996). Morris et al. (1995) described the symptoms of green mold disease as the presence of green fungal sporulation in the mushroom compost or casing layer within 2–5 weeks of the production cycle in the mushroom growing unit. Generally, the crop loss is proportional to the area infected, but no mushrooms are produced in contaminated bags in the case of serious outbreaks. Furthermore, even if fruit bodies appear, they are generally distorted and fetch at very low price or unsalable (Seaby, 1989).

Management:

1. Prevention plays a vital role in green mold management; however, if the infection is present prehand, it has to be controlled. Proper pasteurization of compost and maintenance of good hygienic practices helps to manage the green mold effectively.
2. Among the fungicides evaluated for controlling green mold disease, Environ (a commercial disinfectant) (Abosriwil and Clancy, 2002), Prochloraz, Prochloraz + Carbendazim (Abosriwil and Clancy, 2003) and Thiabendazol (Rinker and Alm, 2008) were the most effective in reducing compost colonization by *Trichoderma* strains. Romaine et al. (2008) recommended imidazole against benzimidazole-resistant strains of *Trichoderma* spp. Abosriwil and Clancy (2003) concluded that the placement of fungicide on the spawn gave better reduction in *Trichoderma* colonization than the application on compost.
3. Biocontrol through application of bacteria like *Bacillus* species that naturally exist in casing are efficient antagonists of aggressive *Trichoderma* strains and can thus be effectively utilized for the management of green mold disease.

10.2.5 FALSE TRUFFLE

Introduction/Economic Importance:

False truffle is considered to be the most dreaded competitor of mushroom fungi. It generally appears when the compost temperature in the bags reach above 22 °C–24 °C.

Casual Organism: *Diehiliomyces microsporus*

Symptoms:

Initially, the mycelium becomes fluffy and turns white in color. Later, the mycelium turns creamy yellow and becomes prominent at the junction of compost and casing layers. Gradually, it gets thicker and solid with appearance of wrinkled mass resembling peeled walnut or brain-like structure.

Their size may range from 0.5 to 3 cm in diameter. At maturity, they turn pink, dry, and reddish and finally disintegrate into a powdery mass.

Management:

1. For effective management of false truffle, it is important that proper hygiene should be observed in and around the mushroom growing facilities. There should be no smell of ammonia in the compost at the time of spawning and pH of compost should be in the range of 7.0–7.5. During cropping, temperature should be kept below 18 °C. However, if the temperature rises above 26 °C–27 °C during spawn run or after casing, it favors truffle. During spawn run, the beds covered with the papers should be moistened twice a week with 0.5% formalin and stumps along with dead mushrooms must be regularly removed from the beds.
2. Spray with Dithane M-45 (0.15%), carbendazim (0.05%) at 8–10 days intervals is found effective. Good cooking out of compost at temperature above 70 °C for 12 h at the end of the crop should be carried out to kill the spores of the pathogen.

10.2.6 BROWN PLASTER MOLD

Introduction/Economic Importance:

Brown plaster mold was first reported on horse dung compost from Missouri. Munjal and Seth (1974) reported the incidence of brown plaster mold from India for the first time and observed severe loses in the white button mushroom.

Casual Organism: *Papulospora byssina*

Symptoms:

Large, dense, and roughly circular patches of mycelium on the casing layer are the characteristic symptoms of brown plaster mold. Initially white, these patches turn brown and powdery with time. The mold can also colonize the compost (Fletcher et al., 1989). Under severe conditions of plaster mold, there remains no mycelium of mushroom.

Management:

1. During composting, care should be taken to add sufficient gypsum with needed amount of water and duration of peak heating should be

of sufficient duration at proper temperatures. The compost should be well fermented and pasteurized.

2. Munjal and Seth (1974) recommended localized treatment of infected patches with 2% formalin, while application of 4% formalin was recommended by Seth and Shandilya (1978).

10.2.7 YELLOW MOLD

Introduction/Economic Importance:

The complex of these fungi produces the yellow mycelial growth in the compost. Yellow mold including mat disease was reported by Kaul et al. (1978). Heavy incidence of yellow mold under seasonal growing conditions in India during the months of February and March causes substantial losses in production of button mushroom.

Casual Organism: *Myceliopthora lutea, Chrysosporium luteum*, and *C. sulphureum*

Symptoms:

Yellowish mycelium of fungal pathogen develops at the interphase of compost and casing layers. Circular colonies of the fungus are formed in the compost. In the initial stages, it is very difficult to identify the disease symptoms as they are not visible outside. The major sources for primary inoculum are air, raw materials used for preparation of compost, and also by mechanical means. Prevalence of more than 70% moisture content leads to severe proportions of the disease.

Management:

1. Proper pasteurization of casing soil and compost should be done to manage the yellow mold in mushroom beds.
2. Strict hygienic conditions should be maintained for the avoidance of pathogen.
3. Appliances and racks used in mushroom cultivation should be properly disinfected with formalin (2%). Spraying of systemic fungicide like Benomyl or Carbendazim is also effective in managing the mold in bed.

10.2.8 WHITE PLASTER MOLD

Introduction/Economic Importance:

This competitor mold is reported to cause severe damage to production and consequent yield of white button mushroom, if the cultivation is done on compost prepared by long method of composting (Kaul et al., 1978; Sohi, 1986; Garcha et al., 1987; Bharadwaj et al., 1989).

Casual Organism: *Scopulariopsis fumicola*

Symptoms:

Dense white patches of mycelium and spores on the casing surface and in the compost are the indicators of occurrence of white plaster mold. These patches give flour like appearance. Unlike other competitor molds, mycelium of *S. fumicola* remains white and does not change its color (Fletcher et al., 1989). This plaster mold appears if the compost at the time of spawning retains smell of ammonia and has pH more than 8.0.

Management:

1. Care should be taken for preparation of good quality compost with proper pasteurization as the fungus reportedly occurs on poor quality compost with high ammonia and pH. Addition of gypsum during the composting is essential to control white plaster mold.
2. Care should be taken to remove ammonia by spreading the prepared compost in air and to maintain proper pH incorporating optimum quantity of compost ingredients.

10.2.9 LIPSTICK MOLD

Casual Organism: *Sporendonema purpurascens*

Symptoms:

The presence of lipstick mold in mushroom crop causes great damage to the crop (Van Griensven, 1988). The mold first appears on the compost during spawn run or on the casing during the cropping period (Howard et al., 1994). The fungus produces a fine, white mycelial growth rather indiscernible from mushroom mycelium (Fletcher et al., 1989). Crystalline colonies of the fungus may resemble frost on a windshield or small white cotton balls on straws or casing. White color of the mycelium gradually changes to bright pink and finally buff, with a powdery appearance at the time of spore production (Fletcher et al., 1989). Dissemination of this fungus generally occurs through spores or any means that transmit them can spread the organism (Howard et al., 1994).

Management:

1. Raising the temperature to 65 °C for 4 h can eliminate lipstick mold, but this also increases the chances of development of other weed molds (Fletcher et al., 1989).
2. Formalin can be used to spot-treat small areas of lipstick mold (Howard et al., 1994) but strict attention to hygiene remains the most effective and safest approach for its control (Fletcher et al., 1989).

10.2.10 PENICILLIUM MOLD/SMOKY MOLD

Casual Organism: *Penicillium* spp.

Symptoms:

Penicillium species are highly opportunistic fungi, which prefer simple carbohydrates but also grow on cellulose, fats, and lignin (Howard et al., 1994). This characteristic green mold is commonly seen in grains of spawn, on trays or the sideboards of shelves or postharvest remnants of mushroom tissue on the casing surface. Apart from the appearance and some inconvenience, this mold has no measurable effect on crop yield or quality (Fletcher et al., 1989).

10.2.11 BLACK WHISKER MOLD

Causal Organism: *Doratomyces microsporus*

Symptoms:

This mold is named due to the production of dark-gray to black whisker-like bristles on the surface of the casing layer (Fletcher et al., 1989). The appearance of this mold in compost is an indicator of unbalanced nutritional base specifically, the presence of certain carbohydrates in the compost at the time of spawning. It has been observed that this mold occurs in concurrence with *Chaetomium* green mold, since both are cellulolytic fungi. Compost heavily infested with this mold appears gray to black because of the high density of spores. When disturbed, the spores are released, resembling smoke. Human allergic responses to the spores have been reported (Howard et al., 1994). The fungus is thought to be antagonistic to mushroom mycelium and may affect the yield of mushrooms (Fletcher et al., 1989).

10.2.12 CINNAMON BROWN MOLD

Causal Organism: *Chromelosporium fulva* (perfect stage—*Peziza ostracoderma*)

Symptoms:

Cinnamon brown mold is found mainly on casing soil just after the application of casing. This mold is firstly noticed as fine, white, aerial mycelium on the compost or casing. Within a few days spores are formed and their color changes from white to light yellow or light golden brown. The spores can be carried along in the air and infect new crops or cause secondary infections (Howard et al., 1994). High air humidity (more than 90%–95%) and a high temperature in the first one and a half weeks after casing are considered to be ideal for the development of cinnamon brown mold (Van Griensven, 1988). The mold can be managed by ensuring that the keeping the relative humidity in the cropping room below 95%. Treatment of casing material with 2% formalin is also very effective. If serious outbreaks of this mold are reported from mushroom farms, this is an indicator of poor hygiene or use of over pasteurized casing material (Howard et al., 1994).

General Management Practices for All Molds:

All plaster molds grow well in compost with a pH of 8.0 or higher and reduce yields by competing with mushroom mycelium (Fletcher et al., 1989; Howard et al., 1994). Although, it is very difficult to manage plaster molds during mushroom cultivation, but the adoption of strict preventive measures may manage its incidence (Van Griensven, 1988). Modification of composting practices to improve compost quality reduces the occurrence of plaster molds.

10.2.13 *INKY CAPS*

Introduction/Economic Importance:

Improper conditioning of the compost leads to the appearance of inky caps during spawn run. The presence of higher levels of nitrogen in ammonical form also results in growth of inky caps.

Causal Organism: *Coprinus* spp.

Symptoms:

Inky caps appear as long slender stalks and thin caps, which later on disintegrate into black slimy mass of spores. Inky caps appear in cream color at first and turns into bluish black at later stages. Inky caps sometimes grow in cluster and have a long sturdy stem impregnated into the compost layer.

Management:

Compost prepared by short method of composting should be properly conditioned during phase-II to remove the ammonical form of nitrogen.

10.3 BACTERIAL DISEASES

Fluorescent Pseudomonas causes several bacterial diseases of cultivated button and oyster mushrooms (Gill, 1995) like *Pseudomonas tolaasii* causing brown blotch disease of *A. bisporus* (Paine, 1919) and the yellowing of *P. ostreatus* (Gill, 1995). However, they are actually complex diseases involving *P. tolaasii* and *P. reactans*. Brown blotch symptoms on white button mushroom are also caused by *P. reactans* (Iacobellis and Cantore, 2003). Mummy disease of *A. bisporus* is known to be caused by *Pseudomonas* spp. (Fletcher et al., 1989) and the drippy gill of *A. bisporus* is known to be caused by *P. agarici* (Young, 1970). However, this bacterium is also reported as the causative agent of the yellowing of *P. ostreatus* (Bessette, 1985) and of the brown discoloration of mushrooms (Lo Cantore and Iacobellis, 2004). Many other bacteria like *Burkolderia gladioli* pv. *agaricicola* causes soft rot and cavity disease of *Agaricus* spp. (Lincon, 1991) and *Ewingella americana* causes internal stipe necrosis of *A. bisporus* (Inglis, 1996). Most common bacterial diseases of mushrooms are summarized as under:

10.3.1 BACTERIAL BLOTCH

Introduction/Economic Importance:

Bacterial blotch disease of *Agaricus bisporus*, caused by the bacterium *Pseudomonas tolaasii* was observed for the first time in the United States of America (Tolaas, 1915) and Paine (1919) defined its etiology. Later on, this disease was also reported from Italy (Ercolani, 1970).

Pseudomonas tolaasii survives in casing material, even after pasteurization. However, disease intensity depends upon the inoculum of bacterial population on the pileus rather than in the casing mixture. Once the disease occurs, the bacteria are spread by splash dispersal during watering, through tools used by pickers and trashers, and by mushroom flies as well as nematodes. It has been observed that if the moisture content of compost is less than 62% at spawning, it preconditions the mushrooms to blotch infection.

Symptoms:

The disease appears as brown, irregular, sunken lesions on the pileus and/or stipe. Under congenial environmental conditions, the lesions that are initially small and separated coalesce and thus affects a large areas of the pileus, which gradually decays with emitting strong offensive smell. Inadequately sterilized soil, contaminated air, and contaminated implements are the main sources of infections. The pathogen spreads through water drops that splash from infected to healthy sporophores. Implements used by pickers, flies, and mites also help in spread of the disease to healthy trays. Circular, yellowish spots develop on the pileus or near the margin and coalesce to form chocolate brown spots, which penetrate the fleshy tissues. In severe cases, brown lesions also appear on the stipe.

Bacterial blotch may develop at any stage of mushroom growth either on the outer surface of sporophore, on stipe or pileus or both. Bacteria that get splashed onto a mushroom surface reproduce under moist conditions, such as on condensation or staying of water drops on the mushroom surface continuously for a number of hours. Bacterial blotch disease is strongly influenced by environment and moisture on the mushroom surface. Therefore, disease control strategies require inhibiting the pathogens reproduction on the mushroom surface. One of the strategies can be cooling the air temperature lower than the desired level followed by reheating a few degrees as this can hold more moisture and ultimately the surface of mushrooms will dry.

Casual Organism: *Pseudomonas tolaasii*

Management:

High humidity (above 85%) and inadequate ventilation during cropping permit the pileus to remain wet for longer periods, which helps in the disease initiation and spread of the disease. Lowering of humidity to 80% and running fans immediately after watering for drying the caps prevent bacteria to spread on the newly emerging mushrooms. Spray of 100 ppm bleaching powder on the beds has also been found very effective.

10.3.2 YELLOWING OF OYSTER MUSHROOMS

Yellowing of oyster mushrooms is caused by *Pseudomonas tolaasii*. This bacterium may infect all the developmental stages of mushroom fruit bodies (Gill, 1995).

If the disease occurs at an early stage, the young sporophores turn yellowish-red in color and show a slow development followed by a rapid wilting. This change may affect the whole bunch or a part. On developed sporophores, yellowish-red lesions that appear depressed and sometimes surrounded by yellow-red halo appear on the pile and/or stipes. Under high temperature and humidity, sporophores rot rapidly and produce an offensive smell. Sometimes, the whole sporophore may get covered by yellow superficial discolored lesions.

Sometimes, the disease may cause significant loss of production, while sometimes only a change in color of sporophores is observed. The disease may be short lived and may affect only the first flush but then disappear and in other occasions the disease may affect the cultivation throughout the production period.

10.4 VIRAL DISEASES

Viruses have been found in association with mushroom fungi and cause several diseases. The causal organisms of La France, watery stipe, dieback, X-disease, or brown disease in mushrooms have been found to be viruses and these diseases may result into severe crop loses. On the basis of size and shape of the associated viral particles, five different viruses have been reported to infect mushrooms that are as follows:

1. MV-1: Spherical particles with a diameter of 25 nm.
2. MV-2: Spherical particles with a diameter of 29 nm.
3. MV-3: Bacilliform particles with 19 × 50 nm.
4. MV-4: Spherical particles with diameter of 35 nm.
5. MV-5: Spherical particles with a diameter of 50 nm.

The above mentioned viruses may infest the crop alone or in any combination with other viruses. Elongation of the stalk with a small, tilted pileus is the most common symptoms of viral disease. Deterioration of the mushroom mycelium is common, which increases with the time resulting into bare patches of the crop. Sometimes small brown mushrooms that often open prematurely develop. Affected fruits bodies are found to be totally water logged when squeezed (watery stipe). Being obligate parasites, these viruses cannot live in soil or outside the host tissues. Hence, mushroom mycelium, spores, spent compost, and affected crop debris are the common source of its survival.

Management:

Maintenance of strict hygiene inside the farm to avoid incidence of viral diseases in mushrooms is very important. Air should be filtered before being used inside the peak heating, spawn running, and cropping rooms. Doors, shutters, racks, and floors should be disinfected with formaldehyde. Mushrooms should be picked before they open. All wooden parts of growing units should be thoroughly cleaned and sterilized to kill any mushroom mycelium from the preceding crop. Strains tolerant or resistant to viruses should be used.

10.5 ABIOTIC DISORDERS

Besides various biotic agents that adversely affect the mushrooms and a large number of abiotic factors that helps to create unfavorable environment for the proper growth of mushrooms often results in the quantitative as well as qualitative loss of mushrooms. Therefore in mushroom cultivation, management of environment is of great significance and any deviation from the optimum requirements may lead to various kinds of abnormalities and problems in growth of mushrooms. Common abiotic agents that create environmental conditions include low or high moisture in the substrate, pH, temperature, relative humidity, CO_2 concentration in the cropping room, wind velocity, fumes, etc. Majority of these factors make the substrate nonselective for mushroom mycelium and encourages the growth of other competitor molds and pests. The common abiotic disorders include storma (sectors or sectoring), weepers (strinkers or leakers), flock (hard cap, open veil, or saggine socks), hollow core and brown pith, scales (crocodiles), purple stem, long stemmed mushrooms, rose comb, and brown discolouration. These are discussed in the sections below.

10.5.1 STROMA

Stroma or sector is a portion of spawn that appears distinctive when compared to the general appearance of spawn. They appear as noticeable aggregations of mushroom mycelium on the surface of spawned compost or on casing. These appear on the compost in small-localized patches, which may coalesce into larger areas and can easily be peeled. On casing, it develops in advance of pinning but rapidly putrefies as soon as watering on the bags is started. Mushrooms may rarely develop on stroma. Stroma formation is related to the genetic character of the spawn but is sometimes induced if spawn is not handled properly or exposed to harmful petroleum based fumes or chemicals or certain detergents during preparation, storage, transit, or at the farm.

10.5.2 WEEPERS/STINKERS/LEAKERS

If considerable amount of water is exuded from the cap of mushrooms they are described as weepers, while they are called as leakers when small water droplets are exuded from the pileus of mushrooms. These water droplets may be sufficiently numerous to cover the mushrooms or few in number and relatively isolated from each other. In the case of leaker, the water droplets remain as droplets and in case of weeper it actually falls or flows. A weeping mushroom can dissolve into white foam. Beneath a weeper, the area develops a putrid odor thus becoming a stinker as the water collects on the casing surface. Although, factors that induce a mushroom to become a weeper are not clear but these are generally observed in beds or bags that contain low-moisture compost (less than 64%) coupled with high moisture casing.

10.5.3 FLOCK

Flock is a physiologically induced malformation of the mushrooms cap and gill tissues. Their cap opens prematurely and the gills of the affected mushrooms are rudimentary that develops poorly with little pigmentation. The flocked mushrooms generally appear during first flush and may disappear in subsequent flushes but in some cases it continues increasing in subsequent flushes. The mechanism that causes the mushrooms to be flocked is genetic. Flock symptoms are also induced by environmental conditions that include diesel exhaust, oil-based point fumes, and certain anticorrosive chemicals in steam boilers or certain diseases like dieback, brown plaster mold, and false truffle.

10.5.4 HARD CAP AND OPEN VEIL

Hard cap is a variation of flock syndrome, wherein the mushroom cap tends to be disproportionately small in relation to stem diameter, while open veil is the premature opening of veil of mushroom with abnormal gill development. Open veil generally occurs when a period of water stress is followed by a generous watering. It also occurs when fumes of certain organic chemicals are released or drift into a growing room. Overall, if open veil appears, it is safe to conclude that the mushrooms have been under stress during their development.

10.5.5 HOLLOW CORE AND BROWN PITH

In case of hollow core of mushrooms, if the bottom of the stem of mushroom is trimmed a circular gap is observed in the center of the stem. The hole may extend the length of the stipe or it may become shorter. When the hollow cut end portion is brown in color, it is called the brown pith. These abnormalities seem to be related to watering and water stress. Sale price of mushrooms is considerably reduced due to these disorders.

10.5.6 PURPLE STEM

It is observed that after harvest or after being in cold storage overnight, cut stems of the mushrooms develop a deep purple color within few hours. At times this color is closer to black than purple. Generally mushrooms from third flush to the end of the crop are most susceptible as polyphenol oxidase, an enzyme that influences pigment formation increases in later flushes. Conditions that predispose mushrooms to this phenomenon are unknown but the frequency and the amount of water applied before harvest seem to affect its occurrence.

10.5.7 ROSE COMB

In case of rose comb, pinkish gills form large lumps and grow on the cap in an abnormal manner giving the appearance of the comb. These abnormal gills may appear at various places on the fruit body giving it a swollen or spongy appearance. These deformed gills often grow at the top of the cap tissue or even on the top of the cap. The mushrooms can even burst or split turning brown making the mushrooms not marketable. This disorder is caused due to gases or vapors coming from solvents, paint, or oil products and polluted casing soil.

10.5.8 SCALES OR CROCODILES

In the cropping room, when the climate is poorly controlled, in particular too much drying out or too great air velocities, it is observed that scales develop on mushrooms, which are manifested in the form of failure of surface tissue to grow and the further development of cap. Strong vapors of formaldehyde or pest-control products in excess can also cause the outer layer of the skin of half-grown mushrooms to tear off. As the mushroom continues to grow, the skin bursts, and so-called crocodile skin is formed.

10.5.9 LONG STEMMED MUSHROOMS

The presence of long stems in combination with a number of other symptoms indicates virus diseases but it is often the result of too high carbon dioxide concentration. Improvement of aeration avoids development of such abnormalities.

10.5.10 BROWN DISCOLORATION

Browning of small pin heads or half-grown mushrooms is commonly observed on seasonal mushroom farms. This may be due to high temperature, sprinkling at high water pressure, spraying with too high chlorine rate, or incorrect use of formalin.

10.6 GENERAL MEASURES TO BE ADOPTED FOR BIOTIC AND ABIOTIC STRESS MANAGEMENT IN MUSHROOMS

There are certain basic practices that can be of immense importance if adopted timely by mushroom growers. Their adoption can save the crop from the havoc of diseases, competitor molds and abiotic disorders. A sum up of the practices are given below:

1. Sanitation and strict hygiene are the most important preventive methods and must be adopted properly. Without them, effective disease or pest control can never be achieved.
2. Every practice should focus on exclusion and elimination of pathogens or pests.
3. Doors of the cropping room should remain closed and practices that expose substrates to pathogens or pests during spawning should be avoided.
4. Screens on windows and doors should be installed to keep mushroom flies away from entering mushroom houses.
5. Regular inspection of mushroom bags or beds should be undertaken carefully for early detection of pests and diseases.
6. Mushroom bags or beds should be cleaned by removing any mushroom debris or stumps shortly after harvest.
7. Floors of cropping rooms should be kept clean. No waste should be dumped near mushroom houses, as this may attract mushroom flies.
8. Spent mushroom substrate must be disinfected or pasteurized before its removal from the mushroom houses.
9. Proper cleaning and thorough disinfection of mushroom houses should be undertaken before raising the next crop.
10. Equipment should be cleaned and disinfected frequently.
11. Entry in mushroom houses should be restricted.
12. Clean clothes and shoes should be worn, and every time before entering the mushroom house hands should be properly washed.

KEYWORDS

- **mushroom**
- **symptomatology**
- **management**
- **ecosystem**
- **biotic factors**

REFERENCES

Abosriwil, S. O. and Clancy, K. J. (2002). A protocol for evaluation of the role of disinfectants in limiting pathogens and weed moulds in commercial mushroom production. *Pest Manag. Sci.* 58: 282–289.

Abosriwil, S.O. and Clancy, K. J. (2003). A mini-bag technique for evaluation of fungicide effects on *Trichoderma* spp. in mushroom compost. *Pest Manag. Sci.* 60: 350–358.

Adie, B., Grogan, H., Archer, S. and Mills, P. (2006). Temporal and spatial dispersal of *Cladobotryum* conidia in the controlled environment of a mushroom growing room. *Appl. Environ. Microbiol.* 72: 7212–7217.

Anderson, M. G., Beyer, D. M. and Wuest, P. J. (2000). Using spawn strain resistance to manage *Trichoderma* green mold. In *Science and Cultivation of Edible fungi: Mushroom Science.* vol. 15, issue 2. pp. 641–644.

Bayer, D. M., Wuest, P. J. and Kremser, J. J. (2000). Evaluation of epidermilolgical factors and mushroom substrate characteristics influencing the occurrence and development of *Trichoderma* green mold. In *Science and Cultivation of Edible fungi: Mushroom Science.* vol. 15, issue 2. pp. 633–640.

Berendsen, Roeland L., Baars, Johan J. P., Kalkhove, Stefanie I. C., Lugones, Luis G., Wösten, Han A. B., Bakker and Peter A. H. M. (2010). *Lecanicillium fungicola:* causal agent of dry bubble disease in white-button mushroom. *Mol. Plant Pathol.* 5: 17–33.

Berendsen, Roeland L., Stephanie, I.C. Kalkhove, Louis G. Lugones, Han A.B. Wösten, and Peter A.H.M. Bakker. (2011). Microbial Inhibition of *Lecanicillium fungicola* in the Mycosphere of *Agaricus bisporus. R. L. Berendsen Thesis.* Utrecht University. pp. 33–51.

Berendsen, Roeland L., Niek Schrier, Stefanie I. C. Kalkhove, Luis G. Lugones, Johan J. P. Baars, Carolien Zijlstra, Marjanne Weerdt, Han A. B. Wosten and Peter A. H. M. Bakker. (2013a). Absence of Induced Resistance in *Agaricus bisporus* against *Lecanicillium fungicola. Antonie Van Leeuwenhoek.* 103(3): 539–550.

Berendsen, Roeland L., Niek Schrier, Stefanie I. C. Kalkhove, Luis G. Lugones, Johan J. P. Baars, Carolien Zijlstra, Marjanne Weerdt, Han A. B. Wosten and Peter A. H. M. Bakker (2013b). Effects of the mushroom-volatile 1-Octen-3-Ol on dry bubble disease. *Appl. Microbiol. Biotech.* 97(12): 5535–5543. DOI:10.1007/s00253–013-4793–1.

Bessette, A. E. (1985). Yellow blotch of *Pleurotus ostreatus. Appl. Environ. Microbiol.* 50: 1535–1537.

Bhatt, N. and Singh, R. P. (2000). Chemical and biological management of major fungal pathogens *Agaricus bisporus* Lange Imbach. *Mushroom Sci.* 15: 587–593.

Bhatt, N. and Singh, R. P. (2002). Chemical control of mycoparasites of button mushroom. *Indian J. Mycol. Plant Pathol.* 32: 38–45.

Chang, S. T. and. Miles, P. G. (1992). Mushroom biology—a new discipline. The Mycologist 6: 64–65.

Clift, A. D. and Shamshad, A. (2009). Modelling mites, molds and mushroom yields in the Australian Mushroom Industry. In: *Proceeding of the 18th World IMACS/MODSIM Congress*, Cairns, Australia, 13–17 July. pp. 491–497.

Cross, M. J. and Jacobs, L. (1969). Some observations in the biology of the spores of the *Verticillium malthousei. Mushroom Sci.* 17: 239–244.

Ercolani, G. L. (1970). Primi risultati di osservazioni sulla maculatura batterica dei funghi coltivati *Agaricus bisporus* (Lange) Imbach.] in Italia: identificazione di *Pseudomonas tolaasii* Paine. *Phytopathol. Medit.* 9: 59–61.

Ferri, F. (1985). *I funghi. Micologia, isolamento, coltivazione.* Bologna: Edagricole., p. 398.

Fletcher, J. T. and Ganney, G.W. (1968). Experiments on the biology and control of *Mycogone perniciosa* Magnus. *Mushroom Sci.* 7: 221–237.

Fletcher, J.T., Jaffe, B., Muthumeenakshi, S., Brown, A. and Wright, D.M. (1995). Variations in isolates of Mycogone perniciosa and in disease symptoms in Agaricus bisporus. *Plant Pathol.* 44(1): 130–140. DOI: 10.1111/j.1365–3059.1995.tb02725.x.

Fletcher, J.T., White, P. F. and Gaze, R. H. (1989). *Mushrooms: Pest and Disease Control.* Andover, UK: Intercept. p. 174.

García-Morrás, J. A. and Oliván, R. (1999). Problemática actual de Trichoderma Pers. In: 2. *Jornadas Técnicas del Champiñón y Otros Hongos Comestibles en Castilla-La Mancha, Casasimarro, Cuenca (España)*, 4–5 Nov 1997, DPC PPE, Cuenca, Spain. pp. 131–140.

Gaze, R.H. (1995). Dactylium or cobweb. *Mushroom J.* 546: 23–24.

Gaze, R. H. (1996). The past year. *Mushroom J.* 552: 24–25.

Gaze, R. H. and Fletcher, J.T. (2008). Mushroom pest and disease control: a color handbook. SD, USA: Academic Press.

Geels, F. P. (1997). Rondetafel- bijeenkomst over *Trichoderma. Champignoncultuur*. 41: 13.

Geels, F. J., van de Geijin and Rutjens, A. (1998). Pests and diseases. In: van Griensven LJLD (Ed.). *The Cultivation of Mushrooms*. East Grinstead, Sussex, England: Interlingua. pp. 361–422.

Gill, W. M. (1995). Bacterial diseases of Agaricus mushrooms. *Rep. Tottori Mycological Inst.* 33: 34–55.

Glamoclija, J., Soković M., Ljaljević-Grbić, M., Vukojević, J., Milenković, I. and van Griensven, L.J.L.D. (2008). Morphological characteristic and mycelia compatibility of different Mycogone perniciosa isolates. *J. Microsc.* 232(3): 489–492. DOI: 10.1111/j. 1365–2818.2008.02145.x.

Godfrey, S. A. C. (2001). Characterization by 16 S rRNA sequence analysis of pseudomonads causing blotch disease of cultivated Agaricus bisporus. *Appl. Environ. Microbiol.* 67: 4316–4323.

Grogan, Helen. (2012). *Detecting Dry Bubble Disease on Mushroom Farms*. Teagasc Technology.

Grogan, H.M. and Gaze, R.H. (2000). Fungicide resistance among *Cladobotryum* spp. causal agents of cobweb disease of the edible mushroom *Agaricus bisporus. Mycol. Res.* 104: 357–364.

Grogan, H., Piasecka, J., Zijlstra, C., Baars, J. J. P. and Kavanagh, K. (2009). *Detection of Verticillium fungicola in Samples from Mushroom Farms using Molecular and Microbiological Methods*. Agricultural Research Forum. p. 131.

Hatvani, L., Antal, Z., Manczinger, L., Szekere, S. A., Druzhinina, I.S., Kubicek, C.P., Nagy, A., Nagy, E., Vagvolgyi, C. and Kredics, L. (2007). Green mold diseases of *Agaricus* and *Pleurotus* are caused by related but phylogenetically different *Trichoderma* species. *Phytopathology.* 97: 532–537.

Hermosa, M. R., Grondona, I. and Monte, E. (1999). Isolation of *Trichoderma harzianum* Th2 from commercial mushroom compost in Spain. *Plant Dis.* 83: 591.

Howard, R. J., Garland, J. A. and Seaman, W. L. (1994). *Mushroom Chapter 26. Diseases and Pests of Vegetable Crops in Canada.* Canada: The Canadian Phytopathological Society.

Hsu, H.K. and Han, Y.S. (1981). Physiological and ecological properties and chemical control of *Mycogone perniciosa* Magnus causing wet bubble in cultivated mushrooms. *Mushroom Sci.* 11: 403–425.

Icobellis, N. S. (2011). Recent advances of bacterial diseases of cultivated mushrooms. In: *Proceedings of the 7th International Conference on Mushroom Biology and Mushroom Products (ICMBMP7).* pp. 452–460.

Iacobellis, N. S. and Lavermicocca, P. (1990). Batteriosi del cardoncello: aspetti eziologici e prospettive di lotta. *Professione Agricoltore.* 2 (2): 32–33.

Icobellis, N. S. and Lo Cantore, P. (2003). *Pseudomonas "reactans"* a new pathogen of cultivated mushrooms. In: (Eds. Iacobellis et al.) *Pseudomonas Syringae Pathovars and Related Pathogens.* Dordrecht, The Netherlands: Kluwer Academic Publishers. pp. 595–605. ISBN-1–4020-1227–6.

Inglis, P.W. (1996). Evidence for the association of the enteric bacterium *Ewingella americana* with internal stipe necrosis of *Agaricus bisporus. Microbiology.* **142**: 3253–3260.

Kaul, T. N., Kachroo, J. L. and Ahmed, N. (1978). Diseases and competitors of mushroom farms in Kashmir valley. *Indian Mushroom Sci.* 1: 193–203.

Kosanović, D., Potočnik, I., Duduk, B., Vukojević, J., Stajić,M., Rekanović, E., & Milijašević-Marčić, S. (2013). *Trichoderma* species on *Agaricus bisporus* farms in Serbia and their biocontrol. *Ann. Appl. Biol.* 163: 218–230.

Kumar, N., Mishra, A. B. and Bharadwaj, M. C. (2014). Effect of *Verticillium fungicola* (PREUSS) HASSEBR inoculation in casing soil and conidial spray on white button mushroom *Agaricus bisporus. Afr. J. Agric. Res.* 9: 1141–1143.

Largeteau-Mamoun, M. L., Mata, G. and Savoie, J. M. (2002). Green mold disease: Adaptation of *Trichoderma harzianum* Th2 to mushroom compost. In: (Eds. Sanchez et al.) *Mushroom Biology and Mushroom Products. Proceedings of The 4th International Conference on Mushroom Biology and Mushroom Products, Cuernavaca, Mexico, 20–22. February 2002.* pp. 179–187.

Lincon, S. P. (1991). Bacterial soft rot of *Agaricus bitorquis. Plant Pathol.* 40: 136–144.

Lo Cantore, P. and Iacobellis, N. S. (2004). First report of brown discolouration of *Agaricus bisporus* caused by *Pseudomonas agarici* in southern Italy. *Phytopathol. Medit.* 43: 35–38.

Manzi, P., Marconi, S., Aguzzi, A. and Pizzoferrato, L. (2004). Commercial mushrooms: nutritional quality and effect of cooking. *Food Chem.* 84: 201–206.

Mattila, P., Suonpaa, K. and Piironen, V (2000). Functional properties of edible mushrooms. *Nutr. Rev.* 16: 694–696.

Mckay, G. J., Egan, D., Morris, E., Scott, C. and Brown, A. E. (1999). Genetic and morphological characterization of *Cladobotryum* species causing cobweb disease of mushrooms. *Appl. Environ. Microbiol.* 65: 606–610.

Morris, E., Doyle, O. and Clancy, K. J. (1995). A profile of *Trichoderma* species. II—Mushroom growing units. *Mushroom Sci.* 14: 619–625.

Munsch, P. (2002). *Pseudomonas costantinii* sp. nov., another causal agent of brown blotch disease, isolated from cultivated mushroom sporophores in Finland. *Int. J. Syst. Evol. Microbiol.* 52: 1973–1983.

Navarro, M. J., Santos, M., Dianez, F., Tello, J. C. and Gea, F. J. (2011). Toxicity of compost tea from spent mushroom substrate and several fungicide *Agaricus bisporus*. In: *Proceedings of the 7th International Conference on Mushroom Biology and Mushroom Products (ICMBMP7)*. vol. 2, pp. 196–201.

Nees, C. G. (1817). *Das system der Pilze and Schwamme, ein Versuch*. Wurzburg: Wuizburg-In der Stahelschen Buchhandlung, Germany, p. 329.

Paine, S.G. (1919). A brown blotch disease of cultivated mushrooms. *Ann. Appl. Biol.* 5: 206–219.

Potocnik, I (2006). Morphological and pathogenic characteristics of the causal agents of dry and wet bubble diseases of white button mushroom (*Agaricus bisporus* (Lange) Imbach) in Serbia. *Pesticidii Fitomedicina*. 21(4): 289–296.

Potocnik, I., Milijasevic, S., Rekanovic, E., Todorovic, B. and Stepanovic, M. (2008). Sensitivity of Verticillium fungicola var. fungicola, Mycogone perniciosa and Cladobotryum spp. to fungicides in Serbia. In: (Ed. M. Van Greuning). *Science and cultivation of edible and medicinal fungi: Mushroom Science XVII, Proceeding of the 17th Congress of the International Society for Mushroom Science*. Cape Town, South Africa: ISMS. pp. 615–627.

Potocnik, I., Vukojevic, J., Stajic, M., Rekanovic, E., Stepanovic, M., Milijasevic, S. and Todorovic, B. (2010a). Toxicity of biofungicide Timorex 66 EC to *Cladobotryum dendroides* and *Agaricus bisporus*. *Crop Protection*. 29: 290–294.

Potocnik, I., Vukojevic, J., Stajic, M., Tanovic, B. and Rekanovic, E. (2010b). Sensitivity of Mycogone perniciosa, pathogen of culinary-medicinal button mushroom *Agaricus bisporus* (J. Lange) Imbach (Agaricomycetideae) to selected fungicides and essential oils. *Int. J. Med. Mushrooms*. 12(1): 91–98.

Potocnik, I., Vukojevic, J., Stajic, M., Tanovic, B. and Todorovic, B. (2008b). Fungicide sensitivity of selected *Verticillium fungicola* isolates from *Agaricus bisporus* farms. *Arch. Biol. Sci.* 60(1): 151–158.

Rinker, D. L. (1996). *Trichoderma* disease: progress toward solutions. *Mushroom World.* **7**: 46–53.

Rinker, D. L. and Alm, G. (2008). Management of casing *Trichoderma* using fungicides. *Mushroom Sci.* 17: 496–509.

Romaine, C. P., Royse, D. J. and Schlagnhaufer, C. (2008). Emergence of benzimidazole-resistant green mould, *Trichoderma aggressivum*, on cultivated *Agaricus bisporus* in North America. *Mushroom Sci.* 17: 510–523.

Romero-Arenas, O., Lara, M. H., Huato, M. A. D., Hernandez, F. D. and Victoria, D. A. A. (2009). The characteristics of *Trichoderma harzianum* as a limiting agent in edible mushrooms. *Revista Colombiana de Biotecnología*. 11: 143–151.

Samuels, G. J., Dodd, S. L., Gams, W., Castlebury, L. A. and Petrini. O. (2002). *Trichoderma* species associated with the green mold epidemic of commercially grown *Agaricus bisporus*. *Mycologia*. 94: 146–170.

Seaby, D. A. (1989). Further observations on *Trichoderma*. *Mushroom*. 197: 147–151.

Seaby, D. A. (1996). Differentiation of *Trichoderma* taxa associated with mushroom production. *Plant Pathology.* 45: 905–912.

Seaby, D. A. (1998). Trichoderma as weed mould or pathogen in mushroom cultivation. In: *Trichoderma and Gliocladium*. (Eds. Harman, G.E., and Kubicek C.P. eds.) *Enzymes,*

Biological Control and Commercial Applications. London: Taylor and Francis. vol. 2, pp. 267–287.

Seth, P. K. (1977). Pathogens and competitors of Agaricus bisporus and their control. *Indian J. Mushrooms.* 3(1): 31–40.

Seth, P. K. and Dar, G. M. (1989). Studies on *Cladobotryum dendroides* (Bukk. Merat) W.Gams et Hozzem, causing cobweb disease of *Agaricus bisporus* (Lange) Singer and its control. *Mush. Sci.* 12(2): 711–723.

Sharma, S. R. (1994). Viruses in mushrooms. In: *Mushroom Biotechnology* (Eds. Nair, M. C., Gokulapalan, C. and Das L.). New Delhi: Indus Publishing Company. pp. 658–685.

Sharma, S. R. and Kumar, S. (2000). Studies on wet bubble disease of white button mushrooms (*Agaricus bisporus*) caused by *Mycogon eperniciosa. Mushroom Sci.* 15: 569–575.

Sharma S. R., Kumar S. 2005. Diseases of mushrooms and their management. (250–251). In: *Challenging Problems in Horticultural and Forest Pathology* (Eds. Sharma R. C. & Sharma J. N).New Delhi: Indus Publishing Company, pp. 246–288.

Sharma, S. R., Kumar, S. and Sharma, V.P. (2002). Diseases and competitor moulds of mushrooms and their management. *J. Early Repub. 22(3): 509–16.*

Sharma, V. P. and Singh, C. (2003). Biology and control of Mycogone perniciosa Magn. Causing wet bubble disease of white button mushroom. *J. Mycol. Plant Pathol.* 33: 257–264.

Sharma, V. P., Suman, B. C. and Guleria, D. S. (1992). *Cladobotryum verticillium* a new pathogen of *A. bitorquis. Indian J. Mycol. Pl. Path.* 22(1): 62–65.

Sinden, J. W. (1971). Ecological control of pathogens and weed molds in mushroom culture. *Annu. Rev. Phytopathol.* 9: 411–432.

Sinden, J. and Hauser, E. (1953). Nature and control of three mildew diseases of mushrooms in America. *Mushroom Sci.* 2: 177–180.

Sobieralski, K., Siwulski, M., Fruzynska-Jozwiak, D. and Gorski, R. (2009). Impact of *Trichoderma aggressivum* f. *europaeum* Th2 on the yielding of *Agaricus bisporus*. *Phytopathologia.* 53: 5–10.

Szczech, M., Staniaszek, M., Habdas, H., Ulinski, Z. and Szymanski, J. (2008). Trichoderma spp.—the cause of green mold on Polish mushroom farms. *Vegetable Crops Res. Bull.* 69: 105–114.

Van Griensven, L. J. L. D. (1988). Wet bubble (*Mycogone perniciosa*). In: *The Cultivation of Mushrooms*. pp. 401–402.

Wells, J. M. (1996). Postharvest discoloration of the cultivated mushrooms *Agaricus bisporus* caused by *Pseudomonas tolaasii*, *P. 'reactans'* and *P. 'gingeri'*. *Phytopathology.* 86: 1098–1104.

Wong, W. C. and Preece, T. F. (1979). Identification of *Pseudomonas tolaasii*: the white line in agar and mushroom tissue block rapid pitting tests. *J. Appl. Bacteriol.* 47: 401–407.

Yang, J. H., Lin, H. C. and Mau, J. L. (2001). Non volatile taste components of several commercial mushrooms. *Food Chem.* 72: 465–471.

Young, J. M. (1970). Drippy gill: a bacterial disease of cultivated mushrooms caused by *Pseudomonas agarici* n. sp. *New Zeal. J. Agr. Res.* 13: 977–990.

CHAPTER 11

DISEASE AND PEST SPECTRUM IN GORGON NUT (*EURYALE FEROX* SALISBURY) CROP AND MANAGEMENT STRATEGY: INDIAN SCENARIO

J. N. SRIVASTAVA[1*], A. K. SINGH[2], and NEERAJ KOTWAL[3]

[1]*Department of Plant Pathology, Bihar Agricultural University, Sabour, Bhagalpur, Bihar, India*

[2]*Division of Plant Pathology, Sher-E-Kashmir University of Agricultural Sciences and Technology (SKUAST-J), Chatha, Jammu 180009, J&K, India*

[3]*Division of Entomology, Sher-e-Kashmir University of Agricultural Sciences and Technology, Jammu, J&K, India*

[*]*Corresponding author. E-mail: j.n.srivastava1971@gmail.com*

ABSTRACT

Gorgon nut *(Euryale ferox* Salisbury) or Makhana *(Euryale Ferox* Salib.) crops are facing several diseases and insect problem in preharvest aquatic condition, which lake a heavy toll of its harvest by minimizing the yield. Makhana is also affected by the diseases like other fruit crops and the damage depends upon the intensity of the disease. Severe diseases that are commonly found on *makhana* crop are seed rot, root rot, and stem rot (*Fusarium* spp., *Pythium* spp., *Phytophthora* spp., and *Rhizoctonia* spp.), Botryis gray mold/gray mold/ Botryis blight (*Botrytis* species), Leaf and floral hypertrophy disease (*Doassansiopsis euryaleae*), Leaf spot disease (*Cercospora* and *Phytophthora* spp.), *Alternaria* leaf blight (*Alternaria tenuis* Nees or *Alternaria alternata*) and heavy population of nematode *Hirschmannielia* were also reported on this crop. In spite of these diseases some other bacterial and nematodes diseases, namely, bacterial rots, etc., because the plants are weak among the insect pests.

11.1 INTRODUCTION

Gorgon nut *(Euryale ferox* Salisbury) is commonly called as Makhana, Fox nut, and Prickly water lily. It is also known as Makhana in Hindi and Bengali, Jewar in Punjabi, Juwar in Kashmeri, Mellunipadmamu in Telugu, Makhanna in Gujrati, Makhano in Marathi, Kuntapadamu in Orria, Thangjing in Manipuri, and Nikori in Assamese (Kak, 1985). It is a flowering plant classified in the water lily family, Nymphaeaceae (Euryalaceae), and grown in stagnant perennial water bodies like ponds, land depressions, oxbow lakes, swamps, and ditches (Thakur, 1978). Gorgon nut/Makhana is a perennial rooted and floating aquatic herb plant. The Gorgon nut/Makhana plant is large prickly with orbicular floating gigantic leaves. The size of Gorgon nut/Makhana plant leaves is even more than 1 m. Gorgon nut/Makhana plant has rooted with rhizomatous stem and cluster of roots. Gorgon nut/Makhana leaves are green above and purple beneath. Gorgon nut/Makhana flowers are violet in color. Gorgon nut/Makhana seeds are also called as Black Diamond (Cronquist, 1981). Gorgon nut/Makhana is a cash crop and marketed in the form of popped Makhana commonly known as Makhana Lawa. Makhana *(Euryale ferox* Salisbury) is native to East Indies and distributed in tropical and subtropical regions of Southeast and East Asia. Gorgon nut/Makhana is cultivated wildly in China from last 3000 years. Reports on the basis of fossil species corroborate that *E ferox* is a temperate plant, introduced through bird dispersal in different parts of the world, mainly India, China, Japan, Korea, North America, Manchuria, Nepal, and Bangladesh. In India, Gorgon nut/ Makhana is grown as a natural crop in stagnant water pool mostly lakes and tanks, in states such as Bihar, Assam, West Bengal, Tripura, Manipur, Orissa, Madhya Pradesh, Rajasthan, J&K, Gzorakhpur in UP, and Alwar in Rajasthan. Out of which Bihar contributes about 80%–85% of gross production of makhana. Around 75% of the total Makhana production comes from Bihar Swetlands. Approximately 2000 tons of popped makhana worth Rs. 100 million are exported outside north Bihar. The major District of Bihar where makhana is grown extensively are Darbhanga, Sitamardhi, Madhubani, Madhepura, Katihar, Purnia, Saharsa, Supaul, Kishanganj, Araria, etc. North Bihar in general and Mithilanchal in particular, abounding in myriad stagnant freshwater pools with repositories of diverse aquaphytes holds unbound potential for cultivation of Makhana crop in this land (Jha et al., 1991b). Although, Gorgon nut/Makhana is cultivated in stagnant water pools/lakes in different parts of the country, the possession of unique skill by the fisherman/farmer community in arduous task of harvesting/collection

"Makhana Guri" or seeds from the deep water beds by making heaps on the beds through countless diverse and processing these seeds into kernels/pops endows this region in commercial cultivation and processing of this crop.

Gorgon nut/Makhana is a good source of carbohydrates, proteins, and minerals. The chemical constituents of the popped kernels in percentages are 12.8% moisture, 76.9% carbohydrate, 9.7% protein, 0.1% fat, 0.5% total minerals, 0.02% calcium, 0.9% phosphorous, and 0.0014% iron (Jha et al., 1991a). According to Shankar et al. (2010), Bilgrami et al. (1983) found makhana superior to dry fruits such as almond, walnut, cashew nut, and coconut in contents of sugar, proteins, ascorbic acid, and phenol (Jha et al. 1991b).

In traditional, Gorgon nut/Makhana markets are popped makhana, which are distinguished into four qualities namely lava, murha, turi, and mix. The differences in popped makhana quality are almost exclusively related with the size of the pop (Prakash and Chaudhary, 1994). Being the noncereal food, Makhana is a food item during the religious fast of human beings. Popped Makhana is used in preparation of a number of delicious and rich sweet dishes, pudding, and milk-based sweets. Apart from being consumed as sweets, it is also used as a thickening object in curries to make it look rich. Roasted for of popped Makhana can also be consumed as a snack preparation. Apart from these, Makhana is used for medicinal purposes as well, both in India and China as documented in the ancient literature. The seed is analgesic and aphrodisiac, hence used in the preparations of a number of Ayurvedic medicines. Makhana is also used as a starch for fabrics and its bran, which is considered as a waste material that can be used in poultry feed.

Similarly to other crops, Makhana crop is badly damaged by at least three major causes namely diseases, insect, and weeds. All these are inflict heavy loss to the crop productivity (Table 11.1).

11.2 DISEASES AND PEST OF GORGON NUT/MAKHANA (EURYALE FEROX SALISBURY) AND THEIR MANAGEMENT

Gorgon nut *(Euryale ferox* Salisbury) or Makhana *(Euryale Ferox* Salib.) crops are facing several diseases and insect problem in preharvest aquatic condition, which lake a heavy toll of its harvest by minimizing the yield. Makhana is also affected by the diseases like other fruit crops and the damage depends upon the intensity of the disease. Severe disease that is commonly found on *makhana* crop are Seed rot, Root rot, and Stem rot (*Fusarium* spp., *Pythium* spp., *Phytophthora* spp., and *Rhizoctonia* spp.), Botryis gray mold/gray mold/Botryis blight (*Botrytis* species), Leaf and floral hypertrophy disease (*Doassansiopsis euryaleae*), Leaf spot disease (*Cercospora* and *Phytophthora* spp.), *Alternaria* leaf blight (*Alternaria tenuis* Nees or *Alternaria alternata*) and heavy population of nematode *Hirschmannielia* were also reported on this crop (Prasad and Haidar, 1968a, 1968b; Haidar and Nath, 1987; Mahto et al., 1993). In spite of these diseases some other bacterial and nematodes diseases, namely, bacterial rots, etc. because the plants are weak. Among the insect pests, Makhana Aphids (*Rhopalosiphum nymphaeae* L.), Makhana beetle (*Donacia delesserti* Guerin-Meneville), Case worm/Moth (*Nymphula crisonalis* Walker), and Snails/Slugs/Gastropods were found to attack the tender leaves and roots during the crop growth (Mishra et al., 1990a, 1990b; Yazdani and Gupta, 1995).

TABLE 11.1 Some Factors that Adversely Affect the Growth and Development of Makhana (*Euryale ferox*) Crop

Duration	Group	Scientific Name
Jan–March	Seed rot, Root rot, and Stem rot	(*Fusarium* spp., *Pythium* spp., *Phytophthora* spp., and *Rhizoctonia* spp.)
	Aphid (Insect)	*Rhopalosiphum nymphaeae*
	Blight organism (Fungus)	*Alternaria alternata*
	Leaf spot disease (Fungus)	(*Cercospora* and *Phytophthora* spp.)
March–May	Case worm/Moth (Insect)	*Nymphula* spp. *(=Elophila)*
	Beetle (stem/root borers Insect)	*Donacia delesserti*
June–August	Leaf, petiole and fruit galls (Fungus)	*Doassansiopsis euryaleae*
February–April	Algal blooms	-
Throughout the crop duration	Water hyacinth	(*Eichhornia crassipes* Mart.)
Throughout the crop duration	*Monocharia* spp.	-

Ponds have diseases and pests just like any other part of the garden. Beset by diseases and pest, the Makhana *(Euryale ferox* Salisbury)/Fox nut/Gorgon nut/Prickly water lily crops may challenge the ability of the gardener to control them with chemicals because chemicals may injure the aquatic life of the pond/pool, the presence of fish is particularly important to the health of pond/pool (Mishra et al., 1990a, 1990b).

Makhana is cultivated in deep and stagnant water pond along with fishes. Hence, may adversely affect the fishes. However, some prophylactic measures and biological methods for the control of diseases is the best. If necessary, choose a brand labeled of chemicals, which is safe for fish and pond life. If fishes are not present in pond you can spray with choice of appropriate chemical for diseases and insects.

Important diseases are mentioned in the sections below.

11.2.1 DISEASES OF MAKHANA (EURYALE FEROX SALISBURY)

11.2.1.1 SEED ROT, SEEDLING ROT, STEM ROT, AND ROOT ROT

Introduction:

Seed rot, Seedling rot, Stem rot, and Root rot diseases of Gorgon nut/Makhana (*Euryale ferox* salib.) are most common and significant disease. Due to the appearance of this disease plant populations may decrease and consequently there are a need of gap filling and also losses in production. There is diversity in pathogens who can cause these diseases, and the most common pathogens were founded associated with these rots, that is, *Fusarium* spp., *Pythium* spp. *Phytophthora* spp., and *Rhizoctonia* spp. They can be reduce germination or cause rot in seeds before germination or cause seedling or root death (Bagyaraj and Govindan, 1996).

Casual Organism: *Fusarium* spp., *Pythium* spp., *Phytophthora* spp., and *Rhizoctonia* spp.

Symptoms:

After the infection by *Fusarium* spp., *Pythium* spp., *Phytophthora* spp., and *Rhizoctonia* spp symptoms appeared as of Seed rot, Seedling rot, Stem rot, and Root rot is usually seen before or immediately after the seedlings emergence. It is very difficult to identify if infection occur prior to emergence and caused damage. Other factors also may be responsible for these problems, so it is important to diagnose very carefully to determine the cause(s). *Fusarium* spp., *Pythium* spp. *Phytophthora* spp., and *Rhizoctonia* spp., maybe cause minutely various symptoms, but similarities can make them difficult to distinguish without laboratory identification (Bagyaraj and Govindan, 1996).

Fusarium spp. is also a major pathogen that infects seeds and seedlings and also infects the tap root system and promotes adventitious root growth near the soil surface due to appearance of light to dark brown lesions on roots (Suryanarayana, 1978). *Fusarium* spp. is typically favored dry and warm soil.

Pythium spp. that infects seeds and seedlings caused rot in seeds or caused preemergence damping-off under moisture conditions. The characteristic symptoms in or on seed and seedling by most *Pythium* spp., infections are soft and weekend seedlings at ground level, which may be brownish in color with water-soaked tissue. This may result in collapsing of the seedlings, which consequently results in their death. Sometimes the symptoms resemble with those of *Phytophthora* spp., in seedlings, and the causal pathogen can only be distinguished after laboratory examination (Suryanarayana, 1978). *Pythium* spp. is typically favored wet and cool soil.

Phytophthora spp. causes seed rot and seedling rot. It produces tan-brown, soft, and rotted tissue. At the primary leaf stage, infected stems appear bruised and soft, the roots appear dark brown or black, which become soft and rotten, the primarily leaves turn yellow, and plants frequently wilt and die (Suryanarayana, 1978). *Ptytophthora* spp. is favored typically wet and warm soil.

Rhizoctonia spp. also infects makhana plants. In seedlings and older plants, a firm, rusty-brown decay or sunken lesion on the root or on the lower stem is a characteristic symptom. The infections can be superficial and cause no noticeable damage, or they can girdle the stem and stunt or kill plants (Suryanarayana, 1978). *Rhizoctonia* spp. is typically favored wet and warm soil.

The favorable conditions for the development of seedling rot and damping off and spread of their pathogens are as follows (Butler, 1918):

1. *Phytophthora* and *Rhizoctonia*—wet and warm soils.
2. *Pythium*—wet and cool soils.
3. *Fusarium*—dry and warm soils.

Gorgon nut/Makhana seed is affected by *Fusarium* spp., *Pythium* spp., *Phytophthora* spp., and *Rhizoctonia* spp. If such seeds sowing in the field gives poor germinate, their root starts with rotting, growth of the plant is checked, and finally plant are dies. The disease also affects the lower parts of vine with small circular to boat-shaped spots, which later on rot and the plant dries up. The disease is seed borne as well as soil borne.

Management:

1. Cleaned the seed and seed coat deeply.

2. Seed treatment with fungicide may reduce seed and seedling rots. If seed treatment with a combination of two or more premixed fungicides are helpful to manage all diseases. For example, products containing mefenoxam/metalaxyl-M @ 2 mL/kg seed or metalayl @ 2.5 g/kg seed may help to reduce damage from fungi such as *Pythium* and *Phytophthora* can be effective against *Pythium* and *Phytophthora*, and products containing fludioxonil (Maxim) @ 2 mL/kg seed or a strobilurin product (azoxystrobin, trifloxystrobin, or pyraclostrobin) @ 2 mL/kg seed can be effective against true fungi such as *Fusarium* and *Rhizoctonia* (Chaube and Varshney, 2003).
3. Cleaned seeds are treated with Emisan 0.25% or Agallol 0.05% or captan 0.25% solution.
4. In the case of soil-borne diseases, the plant residues are collected and burnt and water should be cleaned regularly.

11.2.1.2 BOTRYIS GRAY MOLD/GRAY MOLD/BOTRYIS BLIGHT

Introduction:

Botryis gray mold/Gray mold is fungal disease of Gorgon nut/Makhana. The fungal pathogen attacks on seed and seedlings and also attacks tender parts of the plant in the presence of high humidity. If diseases appeared on seedlings, disease can be seen on leaves and stem. The disease appeared with many characteristics, namely, gray-molded (mycelium of *Botrytis*) or as blight.

Casual Organism: *Botrytis* species

Symptoms:

The symptoms characterized by their first appearance as a white growth on the plant that turns into gray in color and later gives smoky-gray "dusty" appearance due to mass of spore on affected portion of the plant. Severely affected leaves and shoots dieback, and the leaves drop from the plant and fruits also rots (Bagyaraj and Govindan, 1996).

Management:

1. Remove dead or dying tissue from the plants and from the soil surface and destroy them. Sanitation helps in controlling the disease.
2. Before sowing, seeds should be soaked in Hydrogen peroxide and water solution at a ratio of 1:10.

3. Soil application of 20 kg urea and 10 kg of muriate of potash after removal of diseased panicles may be useful for the growth of panicles, which subsequently develop (Agrios, 1988).
4. If no presence of fish in pond water, can be spraying carbendazim/ Thiophanate methyl @0.1% before the onset of cyclonic rains based on weather forecast followed by second spray soon after rains have receded. Applications are needed at intervals of 5 to 7 days in rainy, overcast weather, and every 7–10 days in warm, dry weather (Agrios, 1988).

11.2.1.3 LEAF AND FLORAL HYPERTROPHY DISEASE/LEAF AND FLORAL TUMOR DISEASE OF GORGON NUT/MAKHANA

Introduction:

The observed hypertrophy was on the Gorgon nut/Makhana (*Euryale ferox* salib.) plants growing in North Bihar. This gall/tumor on plant find particularly from Darbhanga (Bihar) in North Bihar (Misra and Mishra, 1973) and further from other parts of India (Mani, 1965, 1973). The causal organism of this hypertrophy was earlier identified and reported as a species of *Synchytrium* by Jha et al. (1997) but after confirmation by International Mycological Institute, London (UK) it is actually a species of *Doassansiopsis,* a member of Ustilaginales. Hence, it has been named as *Doassansiopsis euryaleae* (Verma and Jha, 1999).

Casual Organism: *Doassansiopsis euryaleae* (A Smut Fungus)

Etiology:

Spore balls, a mass of spores, grouped around a central mass of pseudoparenchymatous sterile hyaline cells, were scattered within the enlarged aerenchymatous cavities formed due to hypertrophied cells. They were densely clustered and angularly elliptical to globoid. They ranged from 120 mm × 144 mm to 240 mm × 262 mm (spherical), and 120 mm × 240 mm to 144 mm × 264 mm (oblong) in dia (Agarwal, 1980). The spore balls were surrounded by profusely branched, hyaline, septate, and delicate layer of hyphae (9–15 mm thick). The spores (fertile) were arranged in the outermost layer of the spore balls and were ellipsoidal to prismatic forming palisade layers on the outermost surface. They measured 14–20 mm × 120–140 mm in size and light brown or pale brown. Such elongated spores enclosed a central mass of sterile pseudo-parenchymatous cells measuring 10–33 m each in diameter (Thirumalachar, 1947).

Symptoms:

The crop suffered a heavy damage to its leaves and other parts including the petiole and pedicel. The infection usually extended from leaf lamina to petiole and from pedicel to the basal part of flower causing great distortion in shape due to hypertrophy. At the beginning, the hypertrophy was small but later attained a major dimension. Than yellowish green with purplish margin, irregularly swollen and wrinkled upper laminar portion measuring generally up to 2000 mm^2 in size (50 mm × 40 mm), while in some cases the infection acquires 200 mm × 300 mm area, which makes the infected plant more identifiable. The abnormal thickness of the infected petiole or peduncle makes them more prominent to be spotted in the field only when taken out of the pond. Infection to the basal part of the flower including the ovary greatly reduces the number of viable seeds and thus causes an economic loss to the farmers (Verma and Jha, 2003). Sometimes it affects fruits and causes fruit galls (Thirumalachar, 1947).

The inner aerenchyma of the host plant is highly enlarged due to infection and the cavities are lined with the mycelial mats imparting a white powdery mass in which resting spore balls are profusely lodged. White incrustation within the inner cavities of the infected hypertrophied portion becomes an important anatomical symptom (Jha et al., 1997).

Low water temperature like 17 °C–20 °C is congenial for spread and disease development.

Management:

1. No control measure available/reported for this disease.

11.2.1.4 CERCOSPORA LEAF SPOT DISEASE

Introduction:

Cercospora leaf spot is a common disease of Gorgon nut/Makhana (*Euryale ferox* salib.) similar to other crop plants but it is usually unimportant in well-managed crops.

Casual Organism: *Cercospora* spp.

Symptoms:

Small circular to large irregular shaped spots appear on the leaves. Firstly, the spots remain water soaked and later on turn brown to black in color. Affected leaves turn light yellow and droop, and then the plant dies. Cercospora leaf spot develops rapidly in warm, humid, and wet conditions.

Management:

1. The pond should be cleaned thoroughly and recommended doses of manures and fertilizers should be given timely.
2. Sowing should be done after proper seed treatment.
3. As soon as the symptoms of the disease appear on the plant, Blitox-50 (30 g in 10 L of water) solution should be sprayed provided there is no fish. For every 10 L solution, 10 mL of linseed oil should be mixed before spray.

11.2.1.5 ALTERNARIA LEAF BLIGHT

Introduction:

Prasad and Haidar (1968a, 1968b) first report of *Alternaria tenuis* on Gorgon nut/Makhana (*Euryale ferox*). Jha et al. (1991b) conserve and propagate these taxon seeds were collected from Darbhanga (Bihar) and grown in water tanks in National Botanical Research Institute, Lucknow. In November and December, its leaves developed brown spots, which gradually increased in size causing damage to the whole leaf. The infection was found to be caused by *Alternaria alternata* (Fr.) Keissler. Haidar et al. (1987) have also reported the infection of the same fungus on *Euryale ferox* Salisb. from Bihar. *Alternaria alternate* leaf blight disease is a most serious fungal disease of makhana (Haidar and Nath, 1987; Dwivedi et al., 1995).

Casual Organism: *Alternaria tenuis* Nees or *Alternaria alternata*

Etiology:

In the infected cells, fungal hyphae were both intercellular as well as intracellular causing degradation of cytoplasm in host cells. The fungus induced degradation of cytoplasm in host cells and also the disruption and invagination of host plasma membrane. The cell organelles in the infected cells were migrated and collected on one side toward periphery (Dwivedi et al., 1995).

Frequency of mitochondria was increased and its membrane was disrupted. Several vacuoles of variable size appeared in the cytoplasm of infected cell. Tonoplast was also disrupted and accumulation of osmiophitic particles adjacent to disrupted tonoplast was noticed. Several affected cells showed disorganized collapsed cell organelles and cytoplasm. In such cells, it was difficult to demarcate different organelles. Increased number of mitochondria and disruption of tonoplast in plants infected by pathogen was workers (Gray et al., 1983).

Alternaria alternata was described as a weak parasite. It can exist necrotrophically also. Thus, during infection, initially it attacks the cell, disrupts the plasma membrane, cell wall, and also affects the cellular organelles. Later on, the necrotrophic nature of the fungus results in complete disorganization of the cell.

The effect on ultrastructures of the cell by a pathogen is similar to the effect produced by stress conditions. Under stress conditions, chlorosis and progressive senescence in plants is generally observed.

Symptoms:

Symptoms first appear on the surface of leaf as small irregular reddish-brown flecks with a dark brown or black border. Than it increases in size, spot becomes circular with concentric zone within area affected, spot center become light brown the older portion may fall out leaving a shot-hole appearance. Many spot may coalesce resulting in leaf blight. As leaf spotting increases, blighting and premature defoliation occur (Haider and Mahto, 2003).

Hanchy (1981) also reported changes in chloroplast in some plants infected by different pathogens. According to him, these changes were induced by different toxins. Ten toxin secreted by *Altertmria alternata* causing chlorosis in some plants infected by the fungus was reported by Fulton et al. (1965).

Management:

1. Seed treatment with Captan @0.25% or Thiram @0.25% or Carbendazim @0.1%.
2. Remove infected leaves foliage, and destroy it safely, which will prevent their spread.
3. The best control *of Alternaria tenuis* is to give at least three spray of fytolon @ 15 days interval sprayed provided there is no fish. (Haider and Nath, 1987).

4. Mancozeb @0.25% or Thiophanate methyl @0.25% or Carbendazim @0.1% or chlorothalonil @0.2% or Zineb @0.25% at weekly interval sprayed provided there is no fish.

11.2.1.6 XANTHOMONAS YELLOWING OF LEAF DISEASE

Casual Organism: *Xanthomonas* spp.

Symptoms:

The main symptoms are yellowing and stunted growth of leaves of the Gorgon nut/Makhana plants. Black scars on the stem are also seen. Brownish white fluid is obtained after squeezing the affected stems.

Management:

1. Cleaning of pond water and seed treatment as mentioned earlier should be done.
2. On standing crop, if there is no fish in the pond, 250 ppm Streptocycline or Agrimycin and 0.25%.
3. Blitox-50 solution should be sprayed along with 10 mL of linseed oil per 10 L of spray solution, sprayed provided there is no fish.

Other than the above diseases, Srivastava and Verma (1987) reported that *Chaetomella raphigera* was identified as the cause of a serious leaf spot disease of this aquatic plant grown for its fruits in Manipur, pathogenicity was confirmed experimentally.

11.3 INSECTS OF MAKHANA *(EURYALE FEROX* SALISBURY)

11.3.1 MAKHANA APHIDS (RHOPALOSIPHUM NYMPHAEAE L.)

(Homoptera: Aphididae)

Identification and Distribution:

Rhopalosiphum nymphaeae L. is also called as reddish-brown plum aphid (Patch, 1915). *Rhopalosiphum nymphaeae* is more or less shiny reddish brown to dark olive (brownish on the primary host), and are dusted with a light gray wax, especially on the head, thorax, and legs of immatures. The dorsal cuticle has reticulation formed by regularly shaped roundish bred like spinules. The terminal process of antennal segment 6th is prominently long and about three to four times the length of the base of that segment. The siphunculi are more than twice the length of the tail and are swollen on the distal half. The antennae are about 0.6 times the body length. The tail is elongate and slender. The body length of *Rhopalosiphum nymphaeae* aptera is 1.6–2.6 mm (Stroyan, 1984; Bennett and Buckingham, 2000).

In spring the aphid feeds on various leaf petioles of *Prunus* species and fruit stalks by causing curling of the leaves. In early summer, *Rhopalosiphum nymphaeae* alatae migrates to the secondary hosts such as a large variety of aquatic plants, including Makhana (*Euryale ferox* Salisbury) *Nymphaea* (Water lilies), *Potamogeton* (Pondweeds), Water hyacinth (*Eichhornia crassipes* Mart.) and Water lettuce (*Pistia stratiotes* L.) and *Sparganium* (Arrowheads) and also Rice (*Oryza sativa* L.), numerous cultivated fruit plants (Storey, 2007). The distribution of *Rhopalosiphum nymphaeae* has long been known as a pest of cultivated aquatic plants, transmitting the cauliflower mosaic, abaca mosaic, cabbage black ringspot, cucumber mosaic, and onion yellow dwarf viruses (Gaevskaya, 1969). The rate of development, natality, and survival rates of *Rhopalosiphum nymphaeae* has been studied in relation to its potential for virus transmission both to and aquatic plants and other crops (Ballou et al., 1986).

Biology and Ecology:

This aphid is brown in color in spring and changes to black coloration in early summer they grow wings for their migration. If fruit trees grow near your makhana crops, then examine the trees in spring for colonies of reddish-brown aphids on the undersides of their leaves and stems. A tell-tale sign of the aphids presence is their clear, sticky waste, or honeydew.

Winged adult *Rhopalosiphum nymphaeae* L. migrate from aquatic habitats to fruit trees in late season. The females lay eggs on the trees, and these eggs constitute the overwintering stage. Subsequent generations spend the spring and early summer on the fruit trees. Then, in mid-to-late summer, they migrate to aquatic plants. While they are on fruit trees, they are oviparous (egg laying) but after migrating to hydrophytes, they become ovoviviparous (retain eggs in their oviducts until the eggs hatch and then give birth to living young). The ovoviviparous females are wingless, but those that migrate to and from fruit trees are winged (Stroyan, 1984).

This aphid readily walks on the water surface, often crawling down emergent plant parts to feed underwater. Specialized "hairs" on their bodies trap and hold air while the aphids are underwater. When large numbers of individuals aphid aggregate in a submersed location, the entrapped air bubble sometimes covers the entire colony (Mc Gaha, 1952).

After colonizing aquatic sites, the aphids reproduce quickly, often virtually blanketing the hydrophytes present. The developmental period from the first instar to the adult stage ranges from 7 to 10 days, depending upon optimal temperatures range from 21 °C to 27 °C). Each female produces up to 50 nymphs at an average rate of two to four nymphs per day. The nymphs normally progress through five instars during the course of their development (Scotland, 1940; Mc Gaha, 1952).

Nature of Damage:

Aphids reproduce quickly enough to inflict serious damage, which includes yellowing leaf pads with stunted growth. Aphids are very active which suck the sap from the new leaves of Gorgon nut/Makhana *(Euryale ferox* Salisbury)/Fox nut/Gorgon nut/Prickly water lily. Later on, the leaves become dry and rotten. The affected Gorgon nut/Makhana crop plants remain stunted. The yield is reduced. Most importantly, the insects may infect their alternate hosts with one or more of the five plant viruses they carry. Having these pests on your water lilies means potential trouble throughout the garden (Saraswati et al.,1990).

Management:

1. Control aphids on Gorgon nut/Makhana *(Euryale ferox* Salisbury) are by forcibly spraying with water to dislodge and drown aphids.
2. Light oil sprays will suffocate the aphids and are not harmful to fish or plants. Sprays should be repeated every 10 days to be most effective. Mix two parts of vegetable oil to eight parts of water and a dash of dishwashing detergent. Treat in the evening and rinse off the oil the next morning.
3. If necessary, choose a brand labeled safe for fish of chemicals. Insecticides should not be used as they are toxic to fish and pond life.
4. Ali (2009) found that the Syrphid (*Ischidons cutellaris*) was an especially efficient predator on this aphid. There are also several reports of predation by coccinellids. Saraswati et al. (1990) reported that *Meochilus sexmaculatus*, *Coccinella septempunctata, Micraspis discolor*, *Brumoides suturalis*, and *Scymnus* species controlled populations of *Rhopalosiphum nymphaeae* in Jan–March on *Euyale ferox*, a major cash crop in India. Later in the year, winged forms develop and migrants spread to wide variety of water plants before eventually migrating back to fruit trees, the winter host.

11.3.2 MAKHANA BEETLE (DONACIA DELESSERTI GUERIN-MENEVILLE)

(Coleoptera: Chrysomelidae)

Identification and Distribution:

Donacia delesserti is also called as longhorned leaf beetle. This beetle is brown in color and about one-half inch long. It has a long antenna and long legs. These beetles are also recognized by their thickened upper legs. Adult beetles live on the surface parts of aquatic plants while the larvae feed on submerged stems and roots of Makhana *(Euryale ferox* Salisbury) crops and also 50 aquatic macrophytes (Mac Gillivray, 1903). A few species are limited to a single host, but most feed on several aquatic plants. It is a major pest on Makhana *(Euryale ferox* Salisbury) crops and regularly occurs in Bihar. It is also a pest on water lilies, pickerel, arrowheads, and water cannas (Buckingham et al., 1986; Gaevskaya, 1966).

Biology and Ecology:

Adult beetle is brown in color and about one-half inch long. It has a long antenna and long legs. These beetles are also recognized by their thickened upper legs. Females show great diversity in their manner of oviposition. Some lay their eggs in concentric rows on the undersides of floating leaves. They accomplish this by pushing their abdomens through holes chewed in the leaves. Other species glue egg masses between the overlapping leaves of adjacent plants. Still others insert single eggs directly into the leaf tissue. *Donacia* eggs are oblong in shape, 0.7–1.5 mm long, 0.3–0.7 mm wide, and range from white to yellow in color (Brigham, 1982).

Larvae enclose from the eggs 1–3 weeks after oviposition. Some then migrate to the sediments where they feed on the roots and rhizomes of their host plants. Others feed on submersed leaves and stems (Hoffman, 1940). Larvae have two black spines near the end of their abdomens. These spines reportedly are inserted into plant tissues and serve as a conduit for oxygen. The larval stage is completed in a few weeks for some species but requires nearly a year for others. The latter generally become dormant during colder months and then resume activity as water temperatures warm (Buckingham et al., 1986; Brigham, 1982).

Prior to pupation, the larvae make slits in the roots or submersed stems. Brown, silken cocoons are then constructed over these slits. Oxygen escaping the plants through the slits becomes trapped in the cocoons, thus keeping the pupae well supplied. Duration of the pupal stage is highly variable. Some species overwinter as pupae. Others emerge as adults a few weeks after pupation. Still others are complete development in a few weeks but delay emergence for several months. Most species are believed to live at least 2 years (Wilcox, 1979).

Nature of Damage:

Adult beetles live on the surface parts of aquatic plants while the grubs/larvae feed on submerged stems and roots of Gorgon nut/Makhana (*Euryale ferox* Salisbury) crops where they spend their total life cycle. Newly hatched larvae pierce leaves and stems to acquire air, leaving behind small brown spots referred to as stippling. The insect is active from April to August. As a result of feeding, the whole plant turns yellow, the growth is stunted and ultimately the yield is reduced drastically (Houlihan, 1970; Gaevskaya, 1966).

Management:

1. Once the plants are infested in the pond, it is difficult to control the pest. So, the best way to prevent the pest is to take Gorgon nut/Makhana crop not more than 2–3 years in the same pond.
2. In the absence of fish in ponds apply Phorate granules @ 10 kg/ha or Carbofuran granules @ 30 kg/ha, sprayed provided there is no fish.

11.3.3 CASE WORM/MOTH (NYMPHULA CRISONALIS WALKER)

(Lepidoptera: Pyralidae):

Identification and Distribution:

Adult moths of case worm of *Nymphula crisonalis* are nocturnal and strongly attracted to light. They have irregular white patches on brownish-orange wings. The larvae of plants can change as in cream-colored with brown heads. *Nymphula crisonalis* (Walker) is probably the most recognized aquatic plant insect pest (Sankaran and Rao, 1972; Mangoendihardjo et al., 1977).

Biology and Ecology:

Adult moths of case worm of *Nymphula crisonalis* are 1 inch long and have irregular white patches on brownish-orange wings. Larvae are ¾ inch long and cream-colored with brown heads. Larvae hide in cocoon-like shelters made of leaf pieces joined by silk, hence the nicknames "bagman caterpillars" and sandwich men. Larvae make large holes along aquatic plant leaf margins, and sometimes burrow into leaf stems during their feeding. During late summer, egg clusters are found on the undersides of leaves near the edges, or sometimes in a circle around leaf holes made by beetles (Williams, 1944; Yoshiyasu, 1983; Chakravorty et al., 1977).

Nature of Damage:

This insect is active from March to April. Larvae cause the damage by feeding upon the leaves (Viraktamath et al., 1974). Larvae of *Nymphula crisonalis* hideincocoon-like shelters made of cut leaf of Makhana (*Euryale ferox* Salisbury) crops in pieces joined. Larvae eat large holes along aquatic plant including Makhana (*Euryale ferox* Salisbury) crops leaf margins, and sometimes burrow into leaf stems. The larvas damage the crop, which cut the tender leaves and enter into the case and feed the chlorophyll inside the leaves. As a result, the photosynthesis is adversely affected. Banerji (1972) reported that yearly leaves which are tender more tender and less hairy than the later have been infested by infested by increasingly large number soil lepidopterous larvae, which feed on them and roll them into shelters (Sison, 1938).

Management:

1. Drain-off the water from the crop field of Makhana to kill the floating larvae.
2. Put some kerosene oil in a piece of cloth and dislodge the leaf-cases manually. Collect the cases and destroy them.
3. Spraying the crop with 1.0 L of Quinalphos 25 EC or 1.4 L of Monocrotophos 36SL in 250 L water/ha. if fish not present in ponds. As the pest is nocturnal (active during Night time) in behavior, the spray should be done in the evening hours for getting better results (Banerji, 1972).

11.3.4 SNAILS AND SLUGS OR GASTROPODS: GASTROPODA

Identification and Distribution:

The Gastropoda or gastropods, more commonly known as snails and slugs, are a large taxonomic class Gastropoda and they have all sizes from microscopic to large. Both these animals are amphibious capable on land and water (Kownacka, 1963). Snails differ from slugs in having a spirally coiled shell over their body which in slugs is reduced and completely hidden under mantle. Common garden snail (*Helix* spp. and *Macrochlamysindica*), green house snail (*Opeasgracilis*), and giant African snail (*Achatinafulica*) have been found in different states of India and a lot of damage to our crops. The common Indian slugs are *Limax* spp. *Laevicaulis*alte and other, Arion spp. and *Milax*spp. they thrive well on the leaves and flowers of the wild plants (Rembecka, 1989).

Snails and Slugs or Gastropods have an extraordinary diversification of habitat (Strzelec, 1993). Representatives live in gardens, water bodies like ponds, land depressions, oxbow lakes, swamps, ditches, and other ecological niches, including parasitic ones (Chichester and Getz, 1973).

Biology and Ecology:

All land slugs and snails are hermaphrodites (both male and female reproductive organs are present in the same organism) and have the potential to lay eggs. Mating, egg-laying, hatching, and development are not well synchronized even within a single species, so slugs of various stages of development can be found in a year. This makes slug activity difficult to predict reliably. Adult Snails/Slugs/Gastropods lay spherical, pearly white in color, oval to round eggs in batches of 3–40 beneath leaves, soil cracks, and other protected areas of the plant. They can lay eggs up to six times a year, and for maturation it takes about 2 years. Snails/Slugs/Gastropods reach maturity after about 3–6 months, depending on the species (Barker, 2002, 2004).

Snails and Slugs or Gastropods are most active at night and on cloudy or foggy days. During the day time, they hide under the leaf because of heat and bright light. Often the only clues to their presence are their silvery trails and plant damage. In areas with mild winters, such as southern coastal locations, snails and slugs can be active throughout the year. Snails/Slugs/Gastropods can overwinter in all stages, except in extremely cold winters when adults and juveniles may be killed, but thick snow packs can insulate slugs against the cold. During cold weather, Snails/Slugs/Gastropods hibernate in the topsoil. During hot, dry periods, or when it is cold, snails seal themselves off with a parchment like membrane and often attach themselves to tree trunks, fences, or walls (South, 1992).

Nature of Damage:

Snails and Slugs or Gastropods feed on a various type of living plants and on decaying plant materials. They feed by scraping the surface of their food/ host, which may include seeds, roots, stems, and leaves of plants. They chew irregular holes with smooth edges in leaves and flowers and can be clip succulent plant parts. They also can chew fruit and bark of young plants. Because they prefer succulent foliage or flowers, they primarily are pests of seedlings and herbaceous plants, but they also are serious pests of ripening fruits that are close to the ground such as Gorgon nut/Makhana, strawberries, and some other crops (Dreistadt et al., 1994; Flint, 1998).

Snails and crabs are considered as the most dangerous pest of Gorgon nut/Makhana crops. The snails remain attached to the lower surface of leaves and also feed on them. Some species may attack even the stem and fruits (South, 1992).

Management:

1. They could be handpicked and placing then 10% NaCl solution (South, 1992).
2. The gardener who wishes to eliminate snails may place a piece of cabbage or lettuce in the pool, removing it once after the attraction of snails (Bennet, 1971).
4. Metaldehyde (dust or liquid) 5% dilution has been found to be most effective to control. Since slugs secrete a lot of slime that prevent contact with the irritant, repeated treatment will need in their case (South, 1992).

11.4 NEMATODES INFESTATION ON CROPS OF GORGON NUT/ MAKHANA *(EURYALE FEROX* SALISBURY):

Identification and Distribution:

Nematodes are minute worm-like animal without true body cavity (coelom) and with unsegmented, bilaterally symmetrical, and externally cuticularized body. Nematodes are lower invertebrates, highly diversified and most ubiquitous group of the animal kingdom comprising 80%–90% of all the multicellular animals on the earth. Generally, the nematodes are free living in animals, plants, soil, and also in marine water, freshwater, and stagnant water body. They occur at the bottom of lakes and rivers and at enormous depth in the oceans and in all types of soils. In recent years, there has been a growing realization of the importance of nematodes in causing worldwide loss in field of agricultural crops (Gerber and Smart, 1987).

Nematodes,, namely, *Hirschmanniella, Helicotylenchus, Pratylenchus,* and *Tylenchorhynchus,* found population as parasitic forms and *Chronogaster* found population as nonparasitic form. *Hirschmanniella oryzae* dominated among the parasitic forms and *Chronogaster* dominated among nonparasitic form in rhizosphere of gorgon nut in eutrophic and mesotrophic ponds at growing region (Gerber and Smart, 1987).

Total nematode population as well as population of individual species was found to be much higher in mesotrophic pond than in the eutrophic one. *Tylenchorhynchus* was not present in the eutrophic pond (Mahto et al., 1993; Haidar et al., 1995). Presence of organic matters in eutrophic pond and their decomposition products make the environment less congenial to nematode multiplication resulting in low population density in comparison to the mesotrophic one (Khera and Randhwa, 1985).

Strong seasonal variation in population of *H. oryzae* has been observed in both types of ponds (Haidar et al., 1995). The population remained high during April to October due to favorable temperatures and availability of feeder roots, while during December to June population is low because of unfavorable temperature. Temperature and host plants play important role in population fluctuation of several nematodes (Goheen and Williams, 1955; Gupta and Atwal, 1971; Wallace, 1973).

Nature of Damage:

High population of *H. oryzae* around rhizosphere of gorgon nut and presence of few individuals inside roots suggests that this crop is a good host of *H. oryzae*. Infested roots show discoloration and partially hollow. That root feeding adversely affects absorption of water and nutrients. In lastly, plants show stunted growth in patches.

Management

- Ponds treatment with Carbofuran @ 1.0 kg ai/ha 7 days before sowing/ transplanting.

KEYWORDS

- **gorgon nut**
- **makhana**
- **diseases**
- **symptoms**
- **management**

REFERENCES

Agarwal, D. K. (1980). A new report of *Doassansiopsis hydraphilu* from India. *Indian Phvtopathol.* 33: 338–339.

Agrios, G. N. (1988). *Plant Pathology*. San Diego, New York: Academic Press. p. 803.

Ali, A. (2009). On the predation of aphids by *Ischiodon scutellaris* (Diptera: Syrphidae) under natural environment. *Bionotes.* 11(3): 95–96.

Bagyaraj, D. J. and M. Govindan. (1996). Microbial control of fungal root pathogens. In: (Ed. K. G. Mukerji). *Advances in Botany*. New Delhi: APH Publishing Corporation, pp. 293–321.

Banerji, S. R. (1972). Infestation of *Euryale ferox* Salisb. by larvae of *Nymphula crisonalis* Walker and trials on its control. *J. Bombay Nat. Hist. Soc.* 69: 79–90.

Barker, G. M. (2002). *Molluscs as Crop Pests*. Wallingford, UK: CABI Publishing.

Barker, G. M. (2004). *Natural Enemies of Terrestrial Molluscs*. Oxfordshire, UK: CABI Publishing.

Bennet, G. W. (1971). *Management of Lakes and Ponds*. New York: Van Nostrand, Reinhold Comp. p. 375.

Bennett, C. A. and Buckingham, G. R. (2000). The herbivorous insect fauna of a submersed weed, *Hydrilla verticilla* (Alismatales: Hydrocharitaceae). In: (Ed. Neal R. Spencer) *Proceedings of the X International Symposium on Biological Control of Weeds 4–14 July 1999*, Montana State University, Bozeman, Montana, USA. pp. 307–313.

Bilgrami, K. S., Sinha, K. K. and Singh, A. (1983). Chemical changes in dry fruits during aflatoxin elaboration by *Aspergillus flavus*. *Curr. Sci.* 52(20): 960–964.

Brigham, W. U. (1982). Aquatic Coleoptera. In A.R. Brigham, W.U. Brigham, and A. Gnilka, eds., *Aquatic Insects and Oligochaetes of North and South Carolina*. Midwest Aquatic Enterprises, Mahomet, IL, pp. 10.112–10.128

Buckingham, G. R., Haag, K. H. and Habeck, D. H. (1986). Native insect enemies of aquatic macrophytes: beetles. *Aquatics,* 8(2): 28–34.

Butler, E. J. (1918). *Fungi and Disease in Plants*. Calcutta: Thaker Spink and Co. p. 547.

Chakravorty, S. and Mandal, P. K. (1977). Larval morphology of some leaf eating pests of ice. *Entomon.* 2(2): 201–208.

Chaube, H. S. and S. Varshney. (2003). Management of seed rot and damping-off diseases. In: *Compendium Training Programme of CAS in Plant Pathology*. Pantnagar: GBPUAT. pp. 281–288.

Chichester, L. F. and Getz, L. L. (1973). The terrestrial slugs of Northeastern North America. *Sterkiana.* 51: 11–42.

Cronquist, Arthur (1981). *An Integrated System of Classification of Flowering Plants*. New York: Columbia University Press. p. 111.

Dreistadt, S. H. Clark, J. K. and Flint, M. L. (1994). *Pests of Landscape Trees and Shrubs: An Integrated Pest Management Guide*. Oakland: University California Agriculture Natural Resources. 3359.

Dwivedi, A. K. Shekhar, K. and Sharma, S. C. (1995). Ultrastruciural studies of *Euryale ferax* leaf infected by *Alternaria alternata. Indian Phytopathol.* 48(1): 61–65.

Flint, M. L. (1998). *Pests of the Garden and Small Farm: A Grower's Guide to Using Less Pesticide.* 2nd ed. Oakland: University California Agriculture Natural Resources. 3332.

Fulton, N. D. Bollenbacker, K. and Templeton, G. E. (1965). A metabolite from AHemaria tenuis that inhibits chlorophyll production. *Phytopathology.* 55: 49–51.

Gaevskaya, N. S. (1966). *The Role of Higher Aquatic Plants in the Nutrition of the Animals of Freshwater Basins*. Nauka, Moscow. (Translated by D.G. Maitland Muller. 1969. National Lending Library for Science and Technology, Boston Spa, Yorkshire, England.). p. 159.

Gaevskaya, N. S. (1969). *The Role of Higher Aquatic Plants in the Nutrition of the Animals of Freshwater Basins*. Nauka, Moscow. (Translated by D.G. Maitland Muller. 1969. National Lending Library for Science and Technology, Boston Spa, Yorkshire, England.)

Gerber, K. and Smart, G. C. (1987). Plant parasitic nematodes associated with aquatic vascular plants. In : (Eds. Veech, J. A. and Dickson, D.W.) *Vistas on Nematology*. Hyattsville, Maryland: Society of Nematologists Inc. pp. 488–501.

Goheen, A. C. and Williams, C. F. (1985). Seasonal fluctuation in the population of meadow nematodes in roots of cultivated brambles in North Carolina. *Plant Dis. Rep.* 39(12): 901–905.

Gray, D. J., Amerson, H. V. and Van Dyke, C. G. (1983). Infrastructure of the infection and early colonization of Pinus taeda by Cronanium quercuum formae speciales fusiforme. *Mycologia.* 75: 117–130.

Gupta, J. C. and Atwal, A. S. (1971). Biology and ecology of *Hoplolaimus indicus* (Hoplolaiminae : Nematoda) 11. The influence of various environmental factors and host plants on the reproductive potentials. *Nematologica.* 17(2): 277–284.

Haidar, M. G., Mahto, A. and Mishra, R. K. (1995). *Association of Hirschmaiuniella oryzae* with Euryale ferox and its seasonal fluctuation. *Indian J. Nematol.* 25(1): 115–16.

Haider, M. G. and Mahto, A. (2003). Fungal leaf blight and nematode diseases of Gorgon nut (*Euryale ferox*) and their management. In: (Eds. Mishra, R. K., Jha, Vidyanath and Dharai, P. V.) *Makhana*. New Delhi: ICAR. pp. 159–162.

Haidar, M. G. and Nath, R. P. (1987). Chemical control of *Alternaria* leaf blight of makluma (*Euryale ferox*). *Nat. Acad. Sci. Lett.* 10: 301–302.

Hanchy, P. (1981). Ultrastructural effects. In: (Ed. Durbin R. D.). *Toxins in Plant Diseases*. London, UK: Academic Press Inc. pp. 449–75.

Hoffman, C. E. (1940). Limnological relationships of some northern Michigan Donaciini (Chrysomelidae; Coleoptera). *Trans. Am. Microsc. Soc.* 59: 259–274.

Houlihan, D. F. (1970). Respiration in low-oxygen partial pressures: the adults of Donacia simplex that respire from the roots of aquatic plants. *J. Insect Physiol.* 16: 1607–1622.

Jha, V., Barat, G. K. and Jha, U. N. (1991a). Nutritionan evaluation of *Euryale ferox* Salisb. (Makhana). *Food Sci. Technol.,* **28** (5): 326–328.

Jha, V., Kargupta, A. N., Dutta, R. N, Jha, U. N. Mishra, R. K. and Saraswati, K. C. (1991b) Ultilization and Conservation of *Euryale ferox* Salisbury in Mithila (North Bihar) India. *Aquatic Bot*. 39: 259–314.

Jha, V., Kargupta, A. N., Dutta, R. N., Jha, U.N., Mishra, R.K. and Saraswati, K. C. (1991c). Uttilization and conservation of *Euryale ferox* Salisb. in Mithila (North Bihar), India. *Aquatic Bot.* 9: 295–314.

Jha, V. Sarasvvati, K. C, Kumar, R. and Verma, R. A. B. (1997). First record of a all forming *Synchytrium* species on *E. ferox* Salisbury in India. *Proc. Nat. Acad. Sci.* 62(B) 4: 627–628.

Kak, A.M. (1985). Acquatic and wetland vegetation of the north western himalaya XXI. Family Nympheacea in the North Western Himalaya. *J. Econ. Taxon. Bot.* 7: 591–598.

Khera, S. and Randhawa, N. (1985). Benthic nematodes as indicators of water pollution. *Res. Bull. (Sci.) Punjab Univ*. 36 (3–4): 401–403.

Kownacka, M. (1963). Gastropods of fishponds in Gołysz and Landek (in Polish). *Acta Hydrobiol.* 5: 173–188.

Mac Gillivray, A. D. (1903). Aquatic Chrysomelidae and a table of the families of coleopterous larvae. *Bull. NY. State Museum.* 68: 288–331.

Mahto, A. Haidar, M. G. Mishra, R. K. and Jha, V. (1993). Nematode associations in ponds growing *Euryale ferox Salisbury* (*makhana*) in Darbhanga (North Bihar), India. *J. Freshwater Biol.* 5(4): 299–304.

Mangoendihardjo, S., Setyawati, O., Syedand, R. A., and Sosromarsono, S. (1977). Insects and fungi associated with some aquatic weeds in Indonesia. In: *Proceedings of the 6th Asian-Pacific Weed Scientific Society Conference.* 2. pp. 440–446.

Mani, M. S. (1965). Key to plant galls from India. *Bull. Bot. Surv. India.* 7(1–4): 89–127.

Mani, M. S. (1973). *Plant Galls of India.* India: MacMillan. p. 354.

Mc Gaha, Y. J. (1952). The limnological relations of insects to certain aquatic flowering plants. *Trans. Am. Microsc. Soc.* 71: 355–381.

Mishra, R. K., Jha, V., Kumar, R. and Jha, B. P. (1990a). The pests of Euryale ferox Salisbury in north Bihar. *Environ. Ecol.* 8(1): 133–136.

Mishra, R. K., Jha, V., Kumar, R. and Jha, B. P. (1990b). The pest of Makhana *(Euryale ferox* Salisbury) in North Bihar. *Environ. Ecol.* 8(1):1133–1136.

Misra, A. and Mishra, R. K. (1973). Survey studies of plant galls of Darbhanga, Bihar, India. *Planto.* 3: 44–48

Patch, E.M. (1915). The pond lily aphid as a plum pest. *Science.* 42: 164.

Prakash, O and Choudhary, J. N. (1994). A study on marketing of makhana in Bihar. *Bihar J. Agri. Market.* 2(3): 217–225.

Prasad, H. and Haidar, M. G. (1968a). First report of *Alternaria tenuis* on *makhana* (*Euryale ferox*). *Sci. Cult.* 34: 485.

Prasad, H. and Haidar, M. G. (1968b). First, report of *Alternaria tenuis* on *Makhana* (*Euryale ferox*). *Sci. Cult.* 43: 485.

Rembecka, I. (1989). Freshwater snails (Mollusca: Gastropoda) of fishponds in Silesia (Southern Poland). *Folia Malacol. 3*: 131–137.

Sankaran, T., and Rao, V. P. (1972). An annotated list of insects attacking some terrestrial and aquatic weeds in India, with records of some parasites of the phytophagous insects. *Commonw. Inst. Biol. Contr. Tech. Bull.* 15: 131–157.

Saraswati, K. C., Mishra, R. K., Kumar, R. and Jha, V. (1990). *Rhopalosiphum nymphaeae* (L.) infestation on the leaves of *Euryale ferox. J. Aphidol.* 4(1–2): 89–92.

Scotland, M. B. (1940). Review and summary of insects associated with Lemna minor. *J. NY Entomol. Soc.* 48:319–333.

Shankar, M., Chaudhary, N and Singh, D. (2010). A review on Gorgon nut. *Int. J. Pharm. Biol. Arch.* 1(2): 101–107.

Sison, P. L. (1938). Some observations on the life-history habits and control of the rice case-worm *Nymphula depunctalis* Guen. *Phillipine J. Agric.* 9: 273–301.

South, A. (1992). *Terrestrial Slugs: Biology, Ecology, and Control.* New York, NY: Chapman and Hall. p. 210

Srivastava, L. S. and Verma, R. W. (1987). *Int. J. Trop. Plant Dis.* 7(3): 591–598.

Storey, (2007). Preference and performance of the water lily aphid (*Rhopalosiphum nymphaeae*) among native and invasive duckweeds (Lemnaceae). *M. Biol. thesis.* Georgia Southern University.

Stroyan, H. L. G. (1984). Aphids—Pterocommatinae and Aphidinae (Aphidini). In: *Handbooks for the Identification of British Insects.* London: Royal Entomological Society London. vol. 2, issue 6.

Strzelec, M. (1993). Snails (Gastropoda) of anthropogenic water environments in Silesian Upland (in Poland). *Prace Naukowe Uniwersytetu Śląskiego No. 1358. Katowice*. p. 104.

Suryanarayana, D. (1978). *Seed Pathology*. New Delhi: Vikas Publication.

Thakur, N. K, (1978). Makhana culture. *Indian Farming.* 27: 2327–2329.

Thirumalachar, M. J. (1947). Species of the genera *Doassansia, Doassansiopsis and Burrillia* in India. *Mycologia.* 39: 602–11.

Verma, R. A. B. and Jha, V. (1999). New *Doassansiopsis sp.* associated with *Euryale ferox* Salisb. (*Makluuui*) in north Bihar (India). *J. Freshwater Biol.* 11(1–2): 7–10.

Verma, R. A. B., Jha, V. and Devi, S. (2003). Leaf and floral hypertrophy of Makhana caused by *Doassansiopsis euryaleae*. In: (Eds. Mishra, R. K., Jha, Vidyanath and Dharai, P. V.) *Makhana*. New Delhi: ICAR. pp. 163–168.

Viraktamath, C. A., Puttarudriah, M., and Channabasavanna, G. P. (1974). Studies on the biology of the rice case-worm *Nymphula depunctalis* Guenee (Pyraustidae: Lepidoptera). *Mysore J. Agric. Sti.* 8; 234–241.

Wallace, H. R. (1973). Nematodes as a cause of disease in plants. In : *Nematode Ecology and Plum Disease*. Edward Arnold Publishers Ltd. pp. 146–178.

Wilcox, J. A. (1979). *Leaf Beetle Host Plants in Northeastern North America*. Kinderhook, NY: World Natural History Publications.

Williams, F. X. (1944). Biological studies in Hawaiian water-loving insects. Part 4. Lepidoptera or moths and butterflies. *Proc. Entomol. Soc.* 12: 180–185.

Yazdani, S. S. and Gupta, S. C. (1995). Pest of Makhana and their control. *Indian Farmer Digest.* 28(2–3): 28–29.

Yoshiyasu, Y. (1983). A study of Thailand Nymphulinae (Lepidoptera: Pyralidae) Description of *Paracataclysta* n. gen. *Akitu.* 50: 1–6.

CHAPTER 12

PREVENTION AND MANAGEMENT OF PLANT DISEASES THROUGH PLANT PRODUCTS

J. N. SRIVASTAVA[1] and A. K. SINGH[2]

[1]*Department of Plant Pathology, Bihar Agricultural University, Sabour, Bhagalpur–813210, Bihar, India*

[2]*Division of Plant Pathology, Sher-e-Kashmir University of Agricultural Sciences and Technology, Jammu, India*

Corresponding author. E-mail: j.n.srivastava1971@gmail.com

ABSTRACT

Botanical insecticides have long been touted as attractive alternatives to synthetic chemical insecticides for pest and disease management because botanicals reputedly pose little threat to the environment or to human health. Botanicals are currently used in agriculture in the field of disease management through commercial development of new botanical products. Pyrethrum and neem are well established commercially, pesticides based on plant essential oils have recently entered the market place. Several factors appear to limit the success of botanicals, most notably regulatory barriers and the availability of competing products (newer synthetics, fermentation products, microbials) that are cost-effective and relatively safe compared with their predecessors. In the context of agricultural pest and disease management, botanical insecticides/fungicides are best suited for use in organic food production in industrialized countries but can play a much greater role in the production and postharvest protection of food in developing countries.

12.1 INTRODUCTION

During ancient times, man evolved many cultural and physical practices to protect crops from ravages of insect and pathogens. This was followed by the use of botanicals. The earliest recommendation for the use of plant products was given by Democritus in 470 BC. He recommended the control of plant blights by sprinkling plants with the olive grounds left after extraction of olive oil. Later Surapala in the 9th century suggested the use of vidanga (*Embelia ribis*), mustard and sesame for curing tree diseases in his book "Vrikshayurveda."

As agriculture advanced man invented many new techniques for the control of pathogen. These included the use of inorganic and organic chemicals. The chemical control was so efficient in control of diseases that man started using it indiscriminately. This indiscriminate use continued until 1962 when Rachel Carson for the first time attracted attention toward the bad effect of chemicals in her book "Silent Spring."

Excessive use of chemicals causes soil, air, surface, and groundwater pollution besides affecting the crop produce. Use of chemical pesticides is a not new practice but using for a long time, while that it has many serious drawbacks (Sharaby, 1988), like toxicity against beneficial micro-organisms, insects, fishes, and human (Goodland et al., 1985), pesticide-induced resistance (Georghiou and Taylor, 1977), residual effect that ultimately cause of health hazard (Bhaduri et al., 1989; Pimental et al., 1980). The chemicals have entered our food chain also. Judging the seriousness of this problem, attention has been paid into search for an effective alternative with practically no residual or toxic effects on ecosystem. Botanicals can serve as an alternative approach in this situation.

The natural plant products have been derived from plants, which played a significant role in discovery of a new germicide, nematicide and to certain extent of viricide, either by their direct application to diseases or through their exploitation in the resilienting plant with optimum biological and physical properties. Biopesticides are natural plant products that belong to the so-called secondary metabolites, which include alkaloids, terpenoids, phenolics, and minor secondary chemicals. Plants are rich source of bioactive organic chemicals. It is estimated that the plants may contain as many as 4,000,000 secondary metabolites (Mamun, 2011). According to estimations 2121 plant species have insect-pest control properties. In which some botanicals like neem, ghora-neem, mahogoni, karanja, adathoda, sweet flag, tobacco, derris, annona, smart weed, bar weed, datura, calotropis, bidens, lantana, chrysanthemum, artemisia, marigold, clerodendrum, wild sunflower, and many others. These plants may be grown by farmers at very low cost and extracted by indigenous methods. Plant extract of these plants can be helpful to manage the insect-pest and also safe for environment and human being. Nowdays, botanical pesticides are emerging as one of the prime means to protect crops and their products and prevent environmental pollution from pesticides (Sanjay and Tiku, 2009). Plant extract is easily degradable in comparison to chemical pesticides, so that they are considered as ecofriendly and not toxic to beneficial insect-pests. Most of the botanical pesticides are generally degrade within a few days, and sometimes even within a few hours (Siddiqui and Gulzar, 2003).

This paves the way for the healthy and pollution-free sustainable environment. In prospects of good safety and public health, plant products satisfy the following aims:

1. They are harmless and environmental friendly.
2. Biodegradable.
3. They can increase host metabolism.
4. They are slow in action of limited persistence.
5. Have a shelf life of 2–3 years.
6. Target specific.
7. Low mammalian toxicity.
8. Increase crop quality.
9. Systemic in nature.
10. Optimum bioactivity.
11. Increase benefit: cost ratio.
12. Novel mode of action.
13. Low toxicity.
14. Phytotoxicity of ecotolerance.
15. Less likely development of resistance.
16. Cost efficacy.

However, there are some drawbacks of botanicals control that are as follows:

1. Loss of immediate killing action.
2. Higher dosage requirement.
3. Some plant oils show phytotoxicity of lighter doses, for example, *Azadirachta indica, Cymbopogon* sp.
4. Problems in preparation of formulation of spraying.
5. Some oils are nonedible, so their equipped toxicity problem also needs investigation.

Plant extracts used against plant pathogens and they obtained mainly from trees such as eucalyptus and neem (about 24% of the studies were done with these trees part extracts) and herbaceous plants or plant parts like garlic, citronella, mint, rue, yarrow, ginger, basil, camphor, turmeric, and ocimum (about 54% research work done with these plant extract). Besides these, there is many other plants flora whose antimicrobial potential was tested by researchers. With respect to groups of pathogens, the majority of the work is with those that cause disease in the plant canopy (30% of the works done with these causal agents), like the genus *Alternaria*, *Bipolaris*, *Crinipellis*, *Corynespora*, and *Colletotrichum*. The soil-borne plant pathogens represent 20% of the researches, especially *Rhizoctonia*, *Sclerotium*, *Sclerotinia*, *Fusarium*, and *Phytophthora*. Postharvest pathogens like *Penicillium*, *Aspergillus*, and *Rhizopus* are in 9% of the works and *Meloidogyne* nematode in 9.5%. For the host plants, 30% of the work are with crops like beans, soybeans, coffee, wheat, cotton, and cassava; 20% with vegetables like cucumber and tomato, this later representing alone 15% of all the researches with extract; and 10% with the fruits like papaya, strawberry, and cocoa (Stangarlin et al., 2011)

Some plants and their part used as botanicals:

Common Name	Botanical Name	Family	Plant Part Used
Neem	*Azadirachta indica*	Meliaceae	Leaf, oil cake, bark, fruit pulp.
Garlic	*Allium sativa*	Amaryllidaceae	Cloves
Onion	*Allium cepa*	Amaryllidaceae	Bulbs
Ghrit Kumari	*Aloe vera*	Liliaceae	Leaves
Custered apple	*Annona squamosa*	Annonaceae	Leaves
Satavar	*Asparagus racemosus*	Liliaceae	Root
Kalmegh	*Andrographis paniculata*	Acanthaceae	Leaves, Stem, roots
Bael	*AeglA Eegle mamrelos*	Rutaceae	Leaves and Fruit
Babool	*Acacia nilotica*	Mimosaceae	Leaves, stem, bark
Adusa/vasaka	*Adhatoda vasica*	Acanthaceae	Leaves
Apamarga	*Achyranthus aspersa*	Amranthaceae	Leaves and seeds
Anethium	*Anethum graveolens*	Apiaceae	
Brahmi	*Bacopa monnieri*	Scrophulariaceae	Leaves, whole plant
Palas/Dhak/Tesu	*Butea monosperma*	Leguminaceae	Stem/bark
Baugainvillea	*Bongenvillea spectabilis*	Nyctaginaceae	Leaves
Turmeric	*Curcuma longa*	Zingiberaceae	Leaves and rhizomes

(Table Continued)

Common Name	Botanical Name	Family	Plant Part Used
Amahaldi	*Curcuma amada*	Zingiberaceae	Leaves and rhizomes
Chenopodium	*Chenopodium ambrosioides*	Chenopodiaceae	Leaves
Lemongrass	*Cymbopogon flexuosus*	Poaceae	Leaves
Sadabahar	*Cartharanthus pusillium*	Apocynaceae	Leaves
Citronella	*Cymbopogan winterianus*	Poaceae	Leaves
Cumin	*Cuminum cyminmum*	Apiaceae	Seeds
Periwinkle	*Catharanthus roseus*	Apocynaceae	Leves, flowers
Cannabis	*Cannabis sativa*	Cannabinaceae	Leves, flowers
Papaya	*Carica papaya*	Carcicaceae	Leves, flowers
Amaltas	*Cassia fistula/C. occidentalis*	Caesalpineaceae	Leaves, pods, and seeds
Cleome	*Cleome viscosa*	Capridaceae	Leaves
Coleus	*Colius aromaticus*	Lamiceae	Foliage
Callistemon	*Callistemon lanceolatus*	Myrteaceae	Leaves
Cissus	*Cissus quadrangularis*	Vitaceae	Stem
Datura	*Datura metel, D. stromonium*	Solanaceae	Leaves
Yam	*Dioscoria bulbifera*	Dioscoreaceae	Root tuber rhizomes.
Desmostachya	*Desmostachya bipinnata*	Graminae	Leaves,flower
Eucalyptus	*Eucalyptus citridoa*	Myrtaceae	Leaves and bark
Aonla	*Emblica officinalis*	Euphorbiaceae	Leaves and fruits
Chotidudhi, baridudhi	*Euphorbia microphyla, E. pulcherima*	Euphobiaceae	Leaves
Jatropa	*Jatropa gossypifolia, J. laminarioides*	Euphorbiacceae	Leaves
Bhasmpatti	*Kalanchoe heterophylla*	Labiataeae	Leaves
Lantana	*Lantana camara*	Verbinaceae	Leaves
Pundina	*Mentha arvensis*	Labiateae	Leaves
Murraya/mitha neem	*Murraya konengii*	Rutaceae	Leaves
Melia	*Melia azadirachta*	Meliaceae	Leaves
Shajana	*Moringa olifera*	Moringaceae	Leaves, pods, root, flower
Majorana	*Majorana hortensi*	Labiateae	
Mint	*Mentha* spp.	Labiateae	*Leaves*

(Table Continued)

Common Name	Botanical Name	Family	Plant Part Used
Tulsi	*Ocimum sanctum*	Lambiaceae	*Leaves*
Ocimum	*O. canum giatisimum*	Lambiaceae	*Leaves*
Oxalis	*Oxalis quatosella*	Oxalidaceae	*Leaves*
Solanum	*Solanum nigrum*	Solanaceae	*Leaves, flower*
Sage	*Salvia futicosa*	Labiaceae	*Leaves*
Rumex	*Rumex japonicus*	Polygonaceae	*Leaves*
Nyctanthes	*Nyctanthes arbortistis*	Oleaceae	*Leaves flower*
Marigold	*Tagetues erecta, T. patula*	Compositae	*Whole plant*
Thuja	*Thuja oriantalis*		*Leaves and bark*
Tamrind	*Tamrindus indicus*	Caesalpineaceae	*Seed*
Arjun	*Terminalia arjuna*	Combretaceae	*Leaves, bark*
Fenugreek	*Trigonella foenum granatum*	Papionaceae	*Seeds*
Tiliacora	*Tiliacora racemosa*		*Roots*
Thevitia	*Thevetia peruvina*	Apocynaceae	*Leaves, seed, roots*
Khas	*Viterveria zizanoides*	Poaceae	Roots
Verbina	*Verbena offcinalis*		Leaves
Ashwagandha	*Withamia somnifera*	Solanaceae	Leaves, green berries, seeds, and roots.
Patchauli	*Pogastemon patchaulii*	Labiateae	Leaves
Bachochii	*Psoralea corylifolia*		Leaves, seeds
Carrot grass	*Parthenium hysterophorus*	Astoaceae	Leaves, stems, root, inflorescence
Polyalthia	*Polyalthia longifolia, P. suberosa*	Anonaceae	Leaves
Isabgol	*Plantago ovate*	Plantoginaceae	Seeds, husk
Sarpagandha	*Rauwolfia serpentine*	Apocynaceae	Root, stem, leaves, bark
Castor	*Ricinus communis*	Euphorbiaceae	Seeds and leaves
Kava	*Piper methysticum*	Piperaceae	Roots
Sunflower	*Helianthus annus*	Asteracaea/ Compositae	Receptacle
Hyptis	*Hyptis suveolens*	Lamiaceae	Roots
Clove	*Syzygium aromaticum*	Myrtaceae	Flower buds
Ginger	*Zingibar officinalis*	Zingberaceae	Rhizome
Eelgrass	*Zosetra marina*	Poaceae	Leaves
Sida	*Sida cordifolia*	Malvaceae	Leaves
Ber	*Zizyphus jujube*	Rhamnaceae	Leaves and flower
Nerium	*Nerium oleander*	Apocynaceae	Leaves

12.1.1 ANTIMICROBIAL COMPOUNDS PRESENT IN PLANTS

The presence of antifungal compounds in higher plants has long been recognized as an important factor to diseases resistance (Mahadevan, 1982). Such compounds being biodegradable and selective in their toxicity are considered valuable for controlling some plant diseases (Singh and Dwivedi, 1987). A large number of active secondary metabolic compounds including alkaloids, cyanogenic glycosides, terpenoids, diterpenoids, triterpenoids, lipids, fatty acids, steroids, polyacetylenes, quassinoids, glucosinolates, isothiocyanates, flavonoids, simple and complex phenolics, etc. have reported in many plants, and these secondary metabolites having antimicrobial properties that protect the plant against attack of several pathogens (Chitwood, 2002; Blum, 1996). Some of the important antimicrobial secondary metabolites are as follows:

12.1.1.1 GLYCOSIDES

On wounding of plants, cynogenic glycosides, and glycosinolates and their enzyme mingle and interact producing cyanide, isocyanates, nitriles, and thiocynates. All these compounds are toxic to all organisms and also to fungi (Agrios, 2005)

Detection of Glycosides

One milliliter of water extracts of plants are taken in a test tube and 10 mL of 50% H_2SO_4 added. The mixture is heated in boiling water for 15 min. After this 2 mL of Fehling solution is added and mixture is boiled. Brick red precipitate indicates glycosides.

12.1.1.2 TERPENOIDS

It is a 10 carbon compound. It is a major component of essential plant oils and posseses activity against predators and pathogens.

Detection of Terpenoids (Salkovski test):

A small amount of extract solution is taken and five drops of H_2SO_4 and 1 mL of chloroform is added. Change of yellow color to red indicates the presence of terpenoids.

12.1.1.3 FLAVONOIDES

These are phenolic compounds present in plant tissues especially the vacuoles. They are also present in various bark and heart wood. Flavonoides are inhibitory as well as toxic to pathogens and some of them, for example, mediacarpin, act as phytoallexin in induding defense against fungi (Agrios, 2005).

Detection of Flavonoids (Shinoda Test):

One gram $MgSO_4$ powder and two drops of conc. HCl is added to 3 mL of plant extracts. Red colorations indicate the presence of flavonoides.

12.1.1.4 SAPONINS

Constitutive substances/performed substances made from glycosylated steroidal triterpenoid compounds have antifungal membranolytic activity. Saponins form complexes with cell membrane, form pores, and lead to loss of membrane integrity of pathogens, for example, tomatins in tomato and avenacin in oats.

Detection of Saponins (Frothing Test):

Two milliliters of plant extracts are taken in separate test tubes and are vigourously shaken for 2 min. An observation of frothing indicates presence of saponins.

12.1.1.5 TANNINS

A group of compounds containing phenol, hydroxy acids, and glucosides. They are phenolic oligomers. Condensed tannins are proanthocyanadins and are group of tannin formed by polymerization of flavonoids.

Detection of Tannin:

Two drops of $FeCl_3$ are added to 1 mL extract. A dirty green precipitate showed the presence of tannins.

12.1.1.6 PHENOLS

Toxic compounds accumulate at faster rate after infection, for example, Chlorogenic acid, caffeic acid, and ferulic acid.

Detection of Phenols:

Materials are extracted in ethanol and evaporated to dryness. The residue is then dissolved in distilled water and 0.5 mL Folin Coalteau reagent is added followed by 2 mL of 20% Na_2CO_3 solution. Development of bluish color indicates phenols.

12.1.1.7 ALKALOIDS

Detection of Alkaloids:

Twenty milligrams extract and 2 mL distilled water are taken and filtered. To the filtrate two to four drops of 1% HCl is added and steam is passed through it. Now 1 mL of this solution is mixed with six drops of Wagner's reagent. Brown red precipitate indicates presence of alkaloids.

12.1.2 MANAGEMENT OF FUNGAL DISEASE BY PLANT PRODUCTS

Many types of antifungal compounds are present in plants, which are produced either constitutively or produced in response to attack by pathogen. Compound produced constitutively are called phytoanticipins while those produced in response to attack by pathogen are called phytoallexins.

The constitutive antifungal compounds belong to the following major classes of secondary metabolites:

1. Terpenoids—For example iridoides, sesquiterpines, saponins, etc.
2. Nitrogen- and sulfur-containing compounds— For example alkaloids, amides amines, etc.
3. Aliphatic—long chain alkanes and fatty acids.
4. Aromatics—For example phenol, flavinoids, stilbinoids, biobonyls, xanthones, bimoguimones, etc.

12.1.3 NEEM AND FUNGAL DISEASE MANAGEMENT

Although man is using neem since ancient times for its multifarious uses, its scientific renaissance regarding insecticidal, fungicidal, nematicidal, antiviral properties came into existence quite recently.

Effect of various neem products on different pathogens:

Sr. No.	Neem Product	Effective Against Pathogen
1.	Neem oil cake	*Sclerospora sacchari*
2.	Fruit pulp	*Rhizoctonia solani* (supress formation of sclerotia)
3.	Extract oil	Inhibition of germ tube growth of *Erysiphe polygoni* (Singh and Singh, 1981)
4.	Extract	Mycelial growth and spore germination of *Curvularia lunata, Alternaria alterneta.*
5.	Neem oil	*Fusarium monoliforme, M. phaseolina, D. rostrtatum,* powdery mildew, Black spots.

Neem oil formulation called "Trilogy" has been approved by the EPA for use on food while "Rose Defence" and "Triact" (for control of Powdery mildew, rust, black spot, *Botrytis,* Downy mildew and other common diseases) are designed for use on ornamental.

12.1.4 FUNGAL DISEASES MANAGED BY PLANT PRODUCTS

Name of Fungal Diseases	Causal Agent	Botanical Control	References
Aphanomyces root rot of pea	*Aphanomyces euteiches*	Green manuring with oat, rape and sweet corn, soil application of paper mill residues.	Williums Woodward (1997)
Powdery mildew of pea	*Erysiphe pisi*	Nemadole a neem product, Extracts of *Allium cepa, Allium sativum*, rhizome of ginger and neem.	Singh and Prithviraj (1997)
Anthracnose of bean	*C. lindemuthianum*	Oil of eelgrass	Stanley et al. (2002)
Charcoal rot of soyabean/cluster bean	*Macrophomina phaseolina*	Pearlmillet, residue composet, cauliflower leaf compost	Lodha and Aggarwal (2002)
Web blight of urdbean	*Rhizoctonia solani*	Extract of *Canabis* and *Datura*	Kumar and Tripathi (2012)
Wilt of lentil	*Fusarium oxysporum* f. sp. *lentis*	Extract of ginger, garlic, and neem	Garkoti et al. (2013)

(Table Continued)

Name of Fungal Diseases	Causal Agent	Botanical Control	References
Root rot of French Bean	*Rhizoctonia solani*	Extracts of *Artemisia vulgaris, coix lacryma jobi, Lantana camera*	Mangang and Chhetry (2012)
Wilt of chickpea	*Fusarium oxysporum* f. sp. *ciceri*	Aqueous leaf extracts *Azadirachta indica* and *Lantana camara*	Kamdi et al. (2012)
Root rot of mungbean	*Macrophomina phaseolina*	Oil of palmarosa, lemongrass, citronella, mentha and tulsi	Kumari et al. (2012)
Wilt of pegion pea	*Fusarium oxysporum* f. sp. *udam*	Leaf extract of *Azadirachta indica, Datura festilosa, Tagetes erecta, Eucalyptus citridora, Aegle marmelos* and *Mimusops elengi*	Singh et al. (2010)

12.1.5 VIRAL MANAGED BY PLANT PRODUCTS

Viral diseases are transmitted by many means like mechanical transmission through sap, through vegetative propagating materials, seeds, pollens or through vector like insect, mites, nematodes, and fungi. When virus is transmitted mechanically through sap, intense, sanitation methods are carried out for the control but when it is transmitted by vectors, the best method is to control the vector.

Several antiviral proteins obtained from various plant parts have been found effective in various virus-host system as shown in the below table:

Sr. No.	Antiviral Protein	Source	Virus-Host System
1.	Abrin	*Abrus precatorius*	TMV/*N. glutinosa* seeds
2.	Agrotins	*Agrostemma githago*	TMV/*N. glutinosa* seeds
3.	Bryodin	*Bryonia dioica* leaves and roots	TMV/*N. glutinosa*
4.	Ca-SRI	*Cleodendrum aculeatum* leaves	TMV/*N. glutinosa* TMV/*N. tabaccum/Cymopsis tetragonoloba.*
5.	Dianthins	*Dianthus cryophyllus*	TMV/*N. glutinosa*

Sr. No.	Antiviral Protein	Source	Virus-Host System
6.	EHL(Eranthis Hyemalis Lectin)	*Eranthis hyemalis* bulb	AMV/*Vigna*
7.	Gelonin	*Gelonium multiflorum*	TMV/*N. glutinosa*
8.	MAP	*Mirabilis jalapa* roots	TMV/*N. glutinosa*
9.	Modeccins	*Adenia digitata* roots	TMV/*N. glutinosa*
10.	Momordin	*Momordica charantia*	TMV/*N. glutinosa* seeds
11.	PAP	*Phytolacca americana* seeds leaves	TMV/*N. glutinosa*
12.	:995	*Phaseolous vulgaris*	TMV/*P. vulgaris*
13.	AIMV	*Chenopodium amranticolar*	PVX/*Gmphrema globosa*
14.	RICIN	*Ricinus communis*	ACMV/*N. benthaminana* CaMV/*Brassica compertris*
15.	Soponins	*Saponaria officinalis*	TMV/*N. glutinosa* leaves, root
16.	YLP (yucca recurvi-folia protein)	*Yucca recurvifolia* leaves	TMV/*C. amranticolour*

12.1.6 USE OF BOTANICALS IN VIRAL DISEASE MANAGEMENT OF PULSES

Name of Viral Diseases	Name of Virus	Botanical Control	References
Yellow mosaic disease of mungbean and urdbean	Mungbean yellow mosaic virus	Root extract of *Boerhaavia diffusa* and leaf extract of *Azadirachta indica*	Singh and Awasthi (2009)
Yellow mosaic disease of mungbean	Mungbean yellow mosaic virus	Leaf extract of *Clerodendrum aculeatum,*	Verma and Singh (1994)
Urdbean leaf crinkle	Urdbean leaf crinkle virus (ULCV)	Leaf extract of *Datura metal*	Chaudhury and Saha (1985)
Yellow mosaic of urdbean	Yellow mosaic virus (YMV)	Neem oil and mustard oil	Trivedi et al. (2014)

12.1.7 INDUCED RESISTANCE

Antiviral proteins identified in many plants species can stimulate the active defense mechanism against viruses. Activation of such mechanism results in plants becoming highly resistant to subsequent infection. As for example flower extracts of *Azadirachta indica, Euphorbia milli* induce resistance against PVX infection in hypersensitive host *Chenopodium amranticolor.*

Some plant products and their efficacy:

Sr. No.	Plant Product	Virus	Vectors
1.	Neem oil (5% spray)	RTV	*N. virecens*
2.	Neem cake	RTV	*N. virecens*
3.	Custard apple oil	RTV	*N. virecens*
4.	Custerd + neem oil	RTV	*N. virecens*
5.	Achras sapota	Ragimosaic	*N. virecens*
6.	Basella	Nubra	*N. virecens*
7.	*Mirabilis jalapa*	Nubra	*N. virecens*

12.2 NEMATODE MANAGEMENT BY PLANT PRODUCTS

Phytoparasitic nematodes are very difficult to control because most of them spend their lives in soil or within plant roots and the cuticle and other surface structure of nematodes are impermeable to many organic molecules. Therefore, nematologists are not optimistic about the chemical ways of nematode control. So, attempts are made in the direction of plant products and phytochemical based strategies.

Research with crude plant extract has led to the development of "Sincocin", the trade name of recently developed product containing a mixture of extracts from prickly pear (*Opuntia engelmanni lindheimeri*, Cacteaceae), southern red oak (*Quercus falcota*, Fegaceae), the sumac *Rus aromatica* (Anacardiaceae), and the mangrove *Rhizophora mangle* (Rhizophoraceae). In the field situation, "Sincocin" has been found effective in controlling *Tylenchulus semipenetrans* (citrus nematode) on orange, *Rotylenchulus reniformis* (reniform nematode) on sunflower, and cyst nematode *Heterodera schactti* on sugarbeet. Rizvi et al. (2012) has observed that greatest reduction of *Meloidogyne incognita*, *R. reniformis*, and *T. brassicae* in chickpea by the use of extract of *Argemone mexicana, Calotropis procera, Solanum xanthocarpum*, and *Eichhornia echinulata*. Chickpea plant treated with higher concentrations of leaf extracts of Persian lilac showed the least impact of *M. incognita* (Rehman et al., 2012). Mojumder and Mishra (1993) reported that seed soaking of *Vigna radiata* Linn. with aqueous extracts of Neem reduced the *M. incognita* juveniles in soil. Soil application of Neem and Calotropis leaf extract reduced the root knot nematode infestation in cowpea (Umamaheswari and Sundarababu, 2001a, 2001b). The nematicidal activity of plant extracts may be due to the presence of active toxic principles like azadirachtin and nimbin in *Azadirachta indica* and thiophene and α-terthienyl in *Tagetes erecta*. Such toxic principles might be absorbed by the plant roots, which changes the chemical composition of plants and such roots excerted some influence on the pathogenesis of *M. incognita* as well as improved plant growth (Rao and Parmar, 1984). Due to the exorbitant cost of nematicial chemicals and the environmental hazards they cause, it is high time to modify the nematode management options like botanicals, which are safer and cost-effective.

Mechanism: A number of nematode antagonistic compounds are obtained from many plants. They inhibit or control the nematode population by the following ways:

1. **Inhibition of Hatching of Eggs:** For example isothiocyanates of black mustard seed inhibit hatching of *Globodera rostochenensis* eggs.
2. **Inhibition of Motility:** A combination of linalool and methyl chavicol isolated from essential oil of *Pelargonium graveolens* (Geraniaceae) inhibited motility of *H. cajani* and *M. incognita.*
3. **Inhibiton of Enzyme Action:** All members of Brassicaceae produce thioglucose conjugate called glucosinolates, which are hydrolysed either in soil or by mammal to form isothiocyanates. These isothiocynates react with sulfahydryl group of protein or enzymes and metabolism is disturbed.
4. **Repellant:** For example; Heartwood extract from *Pinus massoniana* (Pinaceae) containing sesquiterpene humulene repelled *B. xylophilus. M. incognita* is repelled by high cucurbitacin produced by cucumber.
5. **Induction of Mortality of Juveniles/Direct Toxicity:** For example; Linalool, geraniol, and citronellol isolated from essential oil of *Pelargonium gravealens* (Geraniaceae) induced mortality in *M. incognita* juveniles. A triterpenoid caumeric acid from *Lantana camara* can directly kill *M. incognita* juvenile at 1% concentration.
6. **Inhibiton of Penetration of Host:** Three Quassinoids from seeds of *Hannoa undulata* (Simarubaceae)-chaparrinone, Klaineanone and glaucaubolone inhibited penetration in tomato roots by *M. incognita* at juvenile stage.

7. **Inhibiton of Host Finding Juvenile Ability:** α-terthienyl from marigold inhibit host finding and induce mortality in entamopathogenic nematode *Steinernema glaseri*.

Some commercial safe and systemic fungicides:

1. Funga stop — Commercial formulation of mint oil
2. Sun Spray — Refined horticultural oil
3. Triology — Neem oil formulation
4. Rose defence — Neem oil formulation
5. Triact — Neem oil formulation
6. Milsana — Knotweed (*Reynoutria sachalinensis*)
7. Spiroxanine
8. Ac-90085
9. 3 preventol
10. 7 L-378257
11. KBR 27- 38 exprodionil.

12.3 CONCLUSION/FUTURE PROSPECT

The pesticidal compounds of plant origin, are much more effective and have little or no side effects on human, still a lot of work needs to be done before large scale utilization of botanical pesticides. Once a potentially useful plant species are identified, intensive breeding and selection work will have to be undertaken so that economic production, from such plant of antagonistic compounds is possible. The safety and selectivity of such botanicals needs to be tested. People are not interested at large in mode of action than in development of environmentally safe, inexpensive agonomicaly useful compounds. The international agronomical industry is well aware of both public perception of chemical hazards and customers deires for alternative methods to protect their liver hood. This has stimulated a wide range of technologist to complement synthetic chemicals with other cost-effective natural products for all types of farmers in future. There is likely that current global thrust on botanical pesticide may yield new leads and prototype modules to guide future research program in plant protection.

KEYWORDS

- **disease management**
- **plant extracts**
- **neem**

REFERENCES

Agrios, G. N. (2005). *Plant Pathology*. 5th edition. London: Elsevier Academic Press. p. 295.

Bhaduri, N., Gupta, D. P. and Ram, S. (1989). Effect of vegetable oils on the ovipositional behaviour of *Callosobruchus chinensis* Fab. In: *Proceedings of the 2nd International Symposium on Bruchids and Legumes (ISBL-2)*. Okayama, Japan. pp. 81–84.

Blum, U. (1996). Allelopathic interactions involving phenolic acids. *J. Nematol.* 28: 259–267.

Chitwood, D. J. (2002). Phytochemical based strategies for nematode control. *Ann. Rev. Phytopathol.* 40: 221–249.

Chowdhury, A. K. and Saha, N. K. (1985). Inhibition of urdbean leaf crinkle virus by different plant extract. *Indian Phytopathol.* 38: 566–568.

Garkoti, A., Kumar, V. and Tripathi, H. S. (2013). Management of vascular wilt of lentil through aqueous plant extracts in tarai region of uttarakhand state. *Bioscan.* 8(2): 473–476.

Georghiou, G. P. and Taylor, C. E. (1977). Pesticide resistance as an evolutionary phenomenon. In: *Proceedings of the 14th International Congress of Entomology*. p759.

Goodland, R., Watson, C. and Ledec, G. (1985). *Biocides Bring Poisoning and Pollution to 3rd World.* The Bangladesh Observer, 16th and 17th January, 1995. P. 3.

Kamdi, D. R., Mondhe, M. K., Jadesha, G., Kshirsagar, D. N. and Thakur, K. D. (2012). Efficacy of botanicals, bio-agents and fungicides against *Fusarium Oxysporum F. Sp. Ciceri*, in chickpea wilt sick plot. *Annals Biol. Res.* 3(11): 5390–5392.

Kumar, S. and Tripathi, H. S. (2012). Evaluation of plant extracts against *Rhizoctonia solani* Kuhn, the incitant of web blight of Urdbean. *Plant Dis. Res.* 27 (2): 190–193.

Kumari, R., Shekhawat, K. S., Gupta, R. and Khokhar, M. K. (2012). Integrated management against rootrot of mungbean [*Vigna radiate* (L.) Wilczek] incited by *Macrophomina phaseolina. J. Plant Pathol. Microb.* 3: 136.

Lodha, S. and Aggarwal, R. K. (2002). Inactivation of *Macrophomina phaseolina* propagule during composting and effect of compost on dry root rot severity and on seed yield of cluster bean. *Eur. J. Plant Pathol.* 108(3): 253.

Mahadevan, A. (1982). *Biochemical Aspects of Plant Disease Resistance Part I. Preferred Inhibitory Substance Prohibition.* New Delhi: Today and Tomorrow Printer and Publishers.

Mamun, M. S. A. (2011). Development of tea science and tea industry in Bangladesh and advances of plant extracts in tea pest management. *Int. J. Sustain. Agr. Tech.* 7(5): 40–46.

Mangang, H. C. and Chhetry, G. K. N. (2012). Antifungal properties of certain plant extracts against *Rhizoctonia solani* causing root rot of French been in organic soils of Manipur. *Int. J. Sci. Res. Publ.* 2(5): 1–4.

Mojumder, V. and Mishra, S. D. (1993). Management of nematode pests. In: *IARI Research Bulletin No. 40: Neem in Agriculture*, (Eds., B.S. Parmar and R.P. Singh), Indian Agricultural Research Institute, New Delhi, pp. 40–48.

Pimental, D., Andow, D., Dyson-Hudson, D., Gallahan, D., Jacobson, S., Irish, M., Croop, S., Moss, A., Schreiner, I., Shepard, M., Thompson, T. and Vinzant, B. (1980). Environmental and social cost of pesticides. A preliminary assessment. *Oikos*. 34: 125– 140.

Rao, K. N. and Parmar, B. S. (1984). A compendium of chemical constituents of neem. Neem News Lett. 1: 39–46.

Rehman, B., Ganai, M. A., Mansoor, K. P. Siddiqui, A. and Usman, A. (2012). management of root knot nematode, Meloidogyne incognita affecting Chickpea (Cicer arietinum) for sustainable production. *Biosci. Int.* 1(1): 01–05.

Rizvi, R., Mahmood, I., Tiyagi, S. A. and Khan Z. (2012). Effect of some botanicals for the management of plant-parasitic nematodes and soil-inhabiting fungi infesting chickpea. *Turk J. Agric*. 36: 710–719.

Sanjay, G. and Tiku, A. K. (2009). Botanicals in pest management current status and future perspectives. *Biomed. Life Sci.* 1: 317.

Sendhilvel, V., Kavitha, K., Nakkeeran, S., Raguchander, T. and Marimuthu, T. (2004). *Glimpses of Plant Pathology*. A. E. Publishers, Coimbatore, p. 120.

Sharaby, A. (1988). Evaluation of some Mytraceae plant leaves as protectants against the infestation by *Sitophilus oryzae* L. and *Sitophilus granarius* L. *Insect Sci. Appl.* 9: 465–468.

Siddiqui, F. A., Gulzar, T. (2003). Tetra cyclic triterpenoids from the leaves ofn*Azadirachta indica* and their insecticidal activities. *Chem. Pharm. Bull.* (Tokyo). (51): 415–417.

Singh, S. and Awasthi, L. P. (2009). Plant products for the management of yellow mosaic disease of mungbean and Urdbean. *J. Plant Prot. Sci.* 1(1): 87–91.

Singh, R. K. and Dwivedi, R. S. (1987). Effects of oils on *Sclerotium rolfsii* causing root rot of barley. *Indian Phytopath.* 40: 531–533.

Singh, P. K., Khan, A., Gogoi, R. and Jaiswal, R. K. (2010). Plant leaf extracts and bioagents for ecofriendly management of wilt of pigeon pea caused by *Fusarium udum. Indian Phytopath.* 63 (3): 343–344.

Singh, U. P. and Prithviraj, B. (1997). Neemazol-a product of neem (*Azadiracta indica*) induces resistance in pea against *Erisiphe pisi. Physiol. Mol. Plant Pathol.* 51: 181.

Stangarlin, J. R., Kuhn, O. J., Assil, L. and Schwan-Estrada K. R. F. (2011). *Control of Plant Diseases using Extracts from Medicinal Plants and Fungi. Science against Microbial Pathogens: Communicating Current Research and Technological Advances*. (Ed. Mandez-Vilas, A.) © Formatex. Formatex Research Center, Badajoz, Spain. pp. 1033–1042.

Stanley, M. S. Callow, M. E., Perry et al. (2002). Inhibition of fungal spore spore adhesion by zosteric acid as the basis for the novel non toxic crop protection technology. *Phytopathology*. 92(4): 378.

Trivedi, A., Sharma, S. K., Ameta, O. P. and Sharma, S. K. (2014). Management of viral and leaf spot disease complexes in organic farming of blackgram. *Indian Phytopath.* 67(1): 97–101.

Umamaheswari, R. and Sundarababu, R. (2001a). Efficacy of neem leaf extract on root knot nematode *Meloidogyne incognita* infecting cowpea. *Indian J. Nematol.* 31(2): 126–128.

Umamaheswari, R. and Sundarababu, R. (2001b). Management of root knot nematode *Meloidogyne incognita* infecting cowpea by Calotropis leaf extract. *Indian J. Nematol.* 31(2): 133–135.

Verma, A. Singh, R. B. (1994). *Clerodendrum aculeatum* a possible prophylactic agent against natural viral infection in mungbean. *Annals Plant Prot. Sci.* 2: 60–63.

Williams-Woodward, J. L. (1997). Green manures of oat, rape and sweet corn for reducing common root rot in pea (*Pisum sativum*) caused by *Aphanomyces euteiches. Plant Soil.* 188(1): 43.

CHAPTER 13

PLANT DISEASE MANAGEMENT THROUGH BIOTECHNOLOGY APPROACHES: AN INSIGHT

SUDHA JALA KOHLI and ASHOK KUMAR THAKUR

P. G. Department of Biotechnology, Tilkamanjhi Bhagalpur University, Bhagalpur 812001, Bihar, India

One of the biggest challenges perpetrated in present times for mankind is that of food security and sustainability, which according to Food and Agricultural Organization has assumed global proportions as per survey reports (http://www.fao.org 2015; http://unionbudget.nic.in/es 2015–2016). According to research, human population, the world over is expected to become around 9.1 billion coupled with a substantial increase in food demand manifested as almost 70% extra by the year 2050 (Godfray et al., 2010; Alexandratos and Bruinsma, 2012). Rapid industrialization has led to environmental issues of deforestation and climate change, land-use mismanagement, crop yield reduction, and postharvest losses due to spoilage, loss of nutrients, and seed viability ultimately resulting in food insecurity (Boxall et al., 2009; Charles et al., 2010; Kumar and Kalita, 2017).

13.1 INTRODUCTION

A comprehensive knowledge and information about plant diseases and their management plays a critical role in maintaining quality and quantity of food grains (Savary et al., 2012a, 2012b). The preharvest and postharvest losses can be attributed to microorganisms (molds, bacteria), insects, rodents, and faulty postharvest management (Mendoza et al., 2016). The problem is exacerbated in humid climates by production of several harmful secondary metabolites known as mycotoxins by diverse phytopathogenic and food spoilage fungi such as *Aspergillus*, *Penicillium*, *Fusarium*, *Cladosporium*, *and Alternaria* (Mendoza et al., 2017). In addition, species such as those belonging to genus *Fusarium*, *Cladosporium*, and *Alternaria* are pathogenic for crop grains during their maturation, whereas *Aspergillus/Penicillium* can infect crops in field but mainly propagate in stored grains (Kabak et al., 2006; Bradford et al., 2018).

Although common agricultural practices such as crop rotation, healthy seed plantation design, soil amendments, and ameliorations in terms of addition of pesticides, fertilizers, and so forth are widely used yet these are time-consuming and labor intensive (Shivayogi et al., 2002). According to data provided by Ministry of Agriculture, 2017–2018 witnessed a damage of agriculture produce worth over 10,000 million dollars due to pests and diseases for almost ten major crops on a national scale A paper entitled "Raising Agricultural Productivity and Making Farming Remunerative for Farmers" published by policy commission of India (NITI Aayog) has highlighted issues related to crop yield losses and possible causes in terms of pathogenesis by several biological organisms such as pests, weeds, insects, nematodes, rodents, and so forth. (niti.gov.in). Agricultural growth has been unsteady in the past few years ranging from 5.8% in 2005–2006 to 0.4% in 2009–2010 and 0.2% in 2014–2015 and 0.8% in 2015–2016 (Deshpande, 2017). In addition, dwindling natural resources, insufficient infrastructure, imbalanced use soil amendments, and so forth have resulted in enhanced complexity of issues related to crop protection and disease management (FAO, 2011; Brown, 2011).

In view of the aforementioned issues, biotechnology has shaped up as a potential tool to deal with issues related to crop diseases and protection based on advanced molecular and genomic tools (Oerke, 2006; Savary et al., 2012a, 2012b). Expressed sequence tags (ESTs), microarrays genetic transformation, and so forth are used to explore genetic basis of stress tolerance for developing stress tolerant crop cultivars (Akshani et al., 2013b). In the recent years, DNA marker technology has helped in development of QTLs mapping, marker aided selection, and genetic transformation to produce superior quality cultivars (Akshani et al., 2013a). Molecular markers can be used for assessment of genetic diversity, fingerprinting genotypes, and separating hybrids from selfed progeny (Haggag, 2008). Gaps in terms of identifying gene and gene function by merging genomics with conventional breeding techniques in order to create novel along with improvised varieties of crops for sustainability in food supply has been followed (Akshani et al., 2015). Process-based agricultural simulation models are helpful for analysis and prediction of crop diseases and their cause and these are helpful in understanding the effect of climate and its factors on various pathosystems (Donatelli et al., 2017).

Genes and gene functions are understood using biotechnological techniques/technologies namely plant tissue culture, genetic engineering, and recombinant DNA technology, which in turn are based on genetic manipulation of living organism plants, animals, microbes, and so forth and signal regulators (Agrios, 2005; Gao et al., 2015). Traditional methods are time consuming and in addition limited availability of genetic resources uncertainty of gene trait stability and subsequent loss of gene pools are issues, which needs redressal (Fagalawa et al., 2013). Molecular technologies are being used to improve plant productivity, product quality, and overall plant health, post removal of constraints caused by pathogens and biotic and abiotic stresses caused by the surrounding environment (Azhaguvel et al., 2006). Multiple-disease resistant crop varieties are being envisaged in terms of strengthening healthy crop yield in agriculture for a sustainable food security worldwide (Fagalawa et al., 2013).

Designing crops resistant to diseases using various biotechnological approaches aims to obtain crops, which display resistance to multiple phytopathgens and are also safe for consumption and environment (Sankarana et al., 2010; Silva et al., 2018). Disease management programs based on pathogen population and interaction with target plants are essential and worked out using methods of disease forecasting or predictive modeling (Zhan et al., 2014; Trivedi et al., 2016). Challenges related to transgenes in terms of choice, origin (i.e. heterologous species, and/or nonhost plants), and gene expression control in terms of signal peptides, gene silencing, and so forth need careful consideration (Silva et al., 2018).

Host-pathogen interaction and the consequential disease cycle depend upon seven fundamental components inoculums produced, dispersal, infection, colonization, symptoms, production of survival structure, and survival (Tiwari et al., 2017). The different stages of a disease manifestation are instrumental in understanding plant disease prediction as well as disease management practice models (De Wolf and Isard, 2007).

Conventionally disease management has been executed in different forms, which include planting diverse crops, fallowing, flooding, mulching, multiple cropping under zero tillage, and multistoreyed cropping (Tiwari et al., 2017). A well-planned routine of use of pesticides and application on standing crops and during postharvest storage is another integral aspect of agriculture (Thurston, 1990). Due to the possible implications on human health and environment attention was given to nonharmful microbes and their products and plant extracts (biopesticides) toward plant disease management (Bhagat et al., 2010, 2014; Tiwari et al., 2017). Several diseases such as rusts, smuts, blights, mildews, and so forth have been effectively treated using these biopesticides (Meena et al., 2013).

13.2 BIOTECHNOLOGICAL APPROACHES—PATHOGENIC BEHAVIOR AND DIVERSITY

A large number of structurally and functionally diverse plant pathogens manifest in the form of mild to extreme symptoms (Strange and Scott, 2005). Host range variability, tissue specificity, optimal environmental conditions, and so forth make it imperative to analyze the biological characteristics of each pathogen followed by a thorough and correct diagnosis culminating in feasible management practices (Klosterman et al., 2016).

13.3 PLANTS AND PATHOGENS—RECOGNITION AND RESPONSE COMMUNICATION

Once a pathogen is recognized by plants, it responds by switching over to a suitable defense mechanism opposite to which pathogens launch a counter attack depending upon its nature, physiology, and metabolism (Silva et al., 2018). Viruses attack plants either mechanically or through a biological vector (insect, nematode, fungus, etc.) into the plant cytosol while bacterial virulent molecules are released into plant cells through secretion systems and fungi attack the apoplast and cytosol of plant cells. Several cell wall degrading enzymes (CWDEs) are secreted by the pathogens, which in turn are counter-acted upon by plants by strengthening cell wall and secretion of CWDE-inhibitors (Sole et al., 2015; Kamber et al., 2017).

Invasion by most of the pathogens is perceived by the plants facilitated by transmembrane plant proteins called pattern recognition receptors (PRRs). These in turn detect microbe-derived molecules termed pathogen-associated molecular patterns (PAMPs). Specific detection of PAMPs by PRRs leads to activation of plants first line of immunity (Uma et al., 2011; Boutrot and Zipfel, 2017; Kachroo et al., 2017). Certain endogenous plant signals are released during invasions called damage-associated molecular patterns, which in turn activate PAMP-triggered immunity (PTI) (Bai et al., 2008). A wide range of effectors or molecules originating from pathogens effecting host cell physiology are released, which results in suppression of PTI via susceptibility proteins (S proteins) leading to host cell infection resulting in effector-triggered susceptibility (ETS) (Chisholm et al., 2006; Jones and Dangi, 2006; Dangi et al., 2013).

Effector production is responded to by plants by developing a second line of resistance in the form of receptors, which are genetically regulated by the so-called resistance (R) genes. These receptors are activated by specific recognition of cognate receptor or pathogen avirulence (Avr) proteins, which give rise to effector-triggered immunity (ETI) (Dangi and Jones, 2006). Across various pathogenic lines PTI involves PAMPs that remain conserved, while ETI is highly specific to some pathogens that secrete unique effector or Avr product. ETI intermittently is characterized by localized programmed cell death, hypersensitive response that restricts spreading of pathogen at the site of infection (Coll et al., 2011). A large number of compounds are secreted by virtue of PTI and ETI to restrict pathogens and infections. These include antimicrobial peptides (AMPs), pathogenesis-related proteins (PR proties) (Candido et al., 2009), ribosome inhibiting proteins (RIPs), and many secondary metabolites (Muthamilarasan and Prasad, 2013; De Ronde et al., 2014; Candido et al., 2015).

Hypersensitive response in infected cells manifests itself in uninfected cells by transferring defense signals through plasmodesmata and to other uninfected sites via phloem resulting in induced distal resistance responses called local acquired resistance and systemic acquired resistance (SAR) (Shah and Zeier, 2013). During pathogenic infection an increase in reactive oxygen species, nitric oxide, and so forth is observed that are triggered by calcium, a ubiquitous intracellular second messenger, which is intrinsically associated with early PTI/ETI perception mechanism (Ma et al., 2008; Seybold et al., 2014; Matika and Loake, 2014). The different layers of innate plant immunity PTI, ETS, and ETI thus induce plant resistance to pathogens and are fine tuned by small noncoding interfering RNA molecules (snciRNA) through distinct yet overlapping genetic and epigenetic RNA interference silencing pathways to silence various genes associated with plant-pathogen reaction (Seo et al., 2013; Weiberg et al., 2014).

Some of the microbial CWDEs have shown good results in terms of enhancing disease resistance in several transgenic plants. These include polygalacturonase-inhibiting proteins (PGIPs) and Xylanase-inhibiting proteins secreted by several pathogenic fungi (Schuttelkopf et al., 2010; Silva et al., 2018). Overexpression of these enzymes in plants such as tomato, tobacco, Arabidopsis, wheat, and so forth resulted in increased resistance to infection by *Botrytis cinerea*, *Rhizoctonia solani*, *Phytopthora capsici*, *Verticillium sp.*, *Fusarium sp.*, and so forth, and microRNA–mediated gene silencing in plant defense and viral counter defense studies have been conducted (Liu et al., 2017a, 2017b; Ma et al., 2015; Khalid et al., 2017). Genetic engineering of disease-resistant plants overexpressing PRRs has been successfully carried out to generate transgenic lines with broad and durable resistance to bacterial and fungal pathogens (Boutrot and Zipfel, 2017).

Mutant recessive S genes (susceptibility) express translation initiation factors 4E and 4G (eIF4E/eIF4G) have been found to impart resistance against *Potyviridae* viruses (Fradin et al., 2011; Schoonbeck et al., 2015). *Arabidopsis thaliana* mutant *eIF4* genes confer resistance to cucumber mosaic virus and turnip crinkle virus (Callot and Gallois, 2014). Using TAL effector technology genome editing-mediated gene knockout was carried out resulting in editing of promoter region of *OsSWEET14* gene, thus inhibiting attachment of *AvrXa7* to S gene promoter thereby inhibiting expression of *OsSWEET14* gene as well (Li et al., 2012; Buttner, 2016)

Plant lectins are known for their crucial role against phytopathogens and their antimicrobial activities have been analyzed in vitro in cell-based assays (Kim et al., 2015). A number of studies have revealed the immunomodulatory and theuraptic properties of lectins (Jandu et al., 2017). Lectins isolated from *Aegle marmelos* have antidiarrheal properties while those from *Punica granatum (PgTe1)* exhibit broad-spectrum antibacterial resistance (Silva et al., 2016). Although these lectins have been tried and tested on animal and human models, advances in protein engineering (Jandu et al., 2017) might help in achieving stress resistant plants based on lectins (Martin and Wang, 2011).

13.4 DEFENSE MECHANISM—AN ARRAY AND ARMY OF BIOMOLECULES

Different resistant proteins of plants represent their pathogen/pest species-specificity along with an inherent versatility in terms of capability of functioning against different types of pathogens. Tomato *Mi gene* (from the CNL group) imparts resistance to root-knot nematodes (*Meloidogyne spp.*), potato aphids (*Macrosiphum euphorbiae*) in addition to sweet potato whitefly (*Bemisia tabaci*) (Bhattarai et al., 2007). Certain antiviral plant R proteins are characterized by corresponding *Avr* molecule, which might be a viral coat protein (CP), replicase, movement protein (MP), or any other viral-encoded factor. *Tomato Sw5* receptor has been observed to perceive tomato spotted wilt virus *NSm MP* (Hallwass et al., 2014), *Arabidopsis* HRT receptor recognizes the *TCV CP* (Ren et al., 2000), and kidney bean RT4-4 receptor senses the *CMV 2a RSS* (Seo et al., 2006).

A number of genes conferring resistance to pathogenic bacteria have been recognized (Dangi et al., 2013; Gururani et al., 2012) with validation in the context of GM plants. Several field trials of GM tomato plants expressing *Bs2*, an R gene from pepper that recognizes *AvrBs2* present in some *Xanthomonas campestris* pathovars, exhibited strong resistance to *Xanthomonas perforans* (Horvath et al., 2012). A number of plant R genes and their cognate fungal *Avr* factors have been elucidated in various pathosystems (Gururani et al., 2012) and these include *Hordeum vulgare-Blumeria graminis*; *Solanum lycopersicum-Fusariumoxysporum*; *Solanum lycopersicum-Cladosporium fulvum*; *Oryza sativa-Magnaporthe grisea*; and *Zea mays-Puccinia sorghi* (Gururani and Park, 2012).

TN13 is a TIR-NBS protein involved in a basal resistance in tomato strains toward infection by *Pseudomonas syringae.* MOS6 is an *Arabidopsis* nuclear transport receptor found to selective interact with TN13 in a transient expression assay and plant defense signaling studies revealed subcellular localization and association of TN13 and MOS6 (Ludke et al., 2018). *Spodoptera littoralis* commonly known as cotton leaf worm is a highly destructive pest of cotton plants. Cotton cultivars were treated with elicitors-methyl jasmonate and sodium nitroprusside and studied for disease resistance at physiological and genetic levels (inter-retrotransposon amplification polymorphism and inter-primer binding site polymorphism retrotransposon-based markers) (Ashry et al., 2018). *Glycine max*(soyabean) infected by charcoal rot ascomycete *Macrophomina phaseolina* showed an increase when parasitized by nematode *Heterodera glycines G. max* [D797-4240 /P1642055] genetic line showing partial resistance to *M. phaseolina*also displayed impairment of *H.glycine* parasitism (Lawaju et al., 2018). *Agrostis stolonifera* was reported to show enhanced disease resistance toward *R. solani* the causative agent for brown patch disease, by foliar application of 2,3-butanediol (2, 3-BD) as a result of Induced Systemic Resistance in grass. In addition, changes in concentrations phytohormones zeatin, abscissic acid, and indole-3-acetic acid, and 3-phenyl propanoid metabolic enzymes—phenyl alaninammonialyase, chalcone isomerase, and 4-Coumarate: Coenzyme A Ligase were observed with best results at 250 mL per litre (Shi et al., 2018).

Various antimicrobial defense proteins of plants encompass AMPs, PR proteins, and RIPs (Narusaka et al., 2014; Silva et al., 2018). These are plant defense peptides synonymous to antibiotics, which work against viruses, bacterial, and fungal infections (Mugford et al., 2009).

AMPs are up to 100 amino acids in length and exhibit structural and functional diversity. Many of these compounds are categorized as defensins, thionins, lipid transfer proteins, snakins, cyclotides, knottins, and hevein-like proteins (Nawrot et al., 2014). In league of transgenic plants resistant to pathogenic bacteria and fungi, wheat thionin expression enhanced disease resistance in *Arabidopsis* against *P. syringae* population and expression of proteins such as cecropin B, phor 21, and so forth against fungi (Oard and Enright, 2006).

A large number of PR proteins accumulate in diseased plants and become a part of defense mechanism of plants (Arlat et al., 1994; Van Loon et al., 2006; Campos et al., 2008). Transgenic potato plants expressing PR-5 osmotin was found to be increasingly resistant to *Phytophthora infestans* along with *Fusarium* sp. and *R. solani* due to differential expression of osmotins and their subsequent role in proline accumulation and as storage reserves (Rivero et al., 2012). Rice chitinase gene subjected to heterologous overexpression in transgenic peanut through *Agrobacterium*-mediated genetic transformation, conferred resistance to *Cercospora arachidicola and Aspergillus flavus* (Prasad et al., 2012). In addition, bacterial chitinase gene obtained from *Streptomyces griseus*in conjunction with plant defensin (AMP and PR-12 protein) from *Wasabia japonica* were co-expressed based on antimicrobial gene pyramiding and were used for transformation of *Solanum tuberosum* plants making them resistant to *Fusarium oxysporum and Alternaria solani* (Khan et al., 2014). Gene pyramiding was used in several studies on transgenic tobacco in which protease inhibitor genes (PR-6 protein) exhibited dual resistance against insects and several phytopathogens (Senthilkumar et al., 2010).

RIPs display N-glycosidase activity on RNA, hydrolyzing a glycosidic bond and removing an adenine residue in a highly conserved sequence at the 3′ terminal region of the ribosomal RNA (rRNA) in the 23/25/28S ribosomal subunits and are also referred to as polynucleotide adenosine glycosidases (Barbieri et al., 1997). Many of these proteins were expected to irreversibly inactivate ribosomes and hamper protein synthesis in targeted phytopathogens (Ng et al., 2010). RIPs exhibit potent in vitro and in vivo antifungal and antiviral activities in transgenic plants by inhibiting protein synthesis (Schrot et al., 2015; Kim et al., 2003; Lanzanova et al., 2009).

Plant innate immunity framework is represented by secondary metabolites, which are also called phytochemicals (Dixon, 2001; Maag et al., 2015) and many owe their origin to several primary metabolites or their biosynthetic intermediates. Many of these compounds are constitutively produced or induced by various environmental stimuli (Piasecka et al., 2015). These are classified as phytoanticipins, constitutively produced and stored in plants and phytoalexins, induced in response to pathogen infection (Piasecka et al., 2015). Among the known phytoanticipins, glucosinolates, cyanogenic glucosides, benzoxazinone glucosides, and saponins have been widely studied (Halkier and Gershenzon, 2006). In crops such as oats saponins have been reported to impart increased resistance to *Gaeumannomyces graminis, F. culmorum, and Fusarium avenaceum* (Bakht et al., 2006; Mugford et al., 2013).

13.5 PATHOGENS DIVERSITY AND COMMONALITY

In plants bacterial species belonging to genus *Erwinia, Agrobacterium, Pseudomonas, Ralstonia, Burkholderia, Acidovorax, Xanthomonas*, and so forth while fungal species of *Ascomycetes and Basidiomycetes* are believed to be pathogenic and proteomic studies of phloem have indicated the possible role of phloem mobile substances or macromolecules as potential tools for SAR (Van Bel and Gaupels, 2004; Lopez-Cobello et al., 2016). The magnitude of crop damage due to diversity of pathogenic fungi makes fungal pathogens interesting causal factors for eliciting plant immune response and synthesis of elicitors (Presti et al., 2015).

Viruses comprise an important and diverse group of pathogens that infect a variety of plants (Prendeville et al., 2014) and the vectors responsible for spread of viruses include insects, mites, nematodes, fungi, and even humans. More than 150 genera of herbaceous dicotyledonous plants including vegetables, flowers, and weeds are infected by tobacco mosaic virus (TMV) (Tepfer, 2002). A number of viruses are responsible for more than 80% loss in crop yield such as Yellow vein mosaic virus, Lettuce infectious yellows crinivirus, and Cucurbit yellow stunting disorder crinivirus (Jose and Usha, 2003; Stewart et al., 2009).

A large number of plant parasitic nematodes derive nutrition from many parts of the plant, including roots, stems, leaves, flowers, and seeds and many are soil borne which feedon root and shoot tissues (Strange and Scott, 2005). A number of parasitic, symbiotic, *cry* protein forming, rhizo bacteria, and several nematophagous bacteria have been explored for their potential as biocontrol agents for controling pathogenic nematodes causing plant diseases (Tian et al., 2007; Nadal et al., 2012). Nematodes such as *Heterodera schachii* and *Meloidogyne incognita* differentially regulate PR protein production and mobilization in target host plants particularly for PR1 to PR5 proteins in roots and leaves of plants like *Arabidopsis thaliana* (Hamamouch et al., 2011). Around 3000 parasitic plant species and 16 families exist, which may follow hemiparasitic and holoparasitic mode of nutrition (Musselman and Press, 1995). Parasitic seed plants, such as dodder (Cuscuta), belonging to family Cuscutaceae, depend entirely on their host for their existence (Kaiser et al., 2015). Knowledge of metabolic networks and identification of specific targets (genes) is essential for study of plant pathogenesis (Quieter, 2016).

In addition to identification of genes responsible for pathogenicity, the resistant genes also need to be essentially identified (Beljah et al., 2015). Plant pathogens can be biotrophic (feed on living tissue), necrotrophic (feed on dead tissue), and many are hemibiotrophic (Okmen and Doehlemann, 2014; Mendez and Romera, 2017)

13.6 WHY GENETIC RESISTANCE?

Studies on genetic basis of disease resistance offer an effective approach for plant disease management. Resistant genes may be clustered or may occur in tandem repeats, implying different specificities for resistant genes that might be attributed to gene duplication followed by intragenic and intergenic recombination, gene conversion, and diversifying selection (Vincelli, 2016). In a broader perspective genetic engineering techniques involving genetic recombination and targeted mutagenesis have not been sufficient to cope with disease resistance in crops owing to increasing food demand thus paving way for genomics-based transgenesis based on use of CRISPR-associated proteins CRISPR-Cas technology (Clustered Regularly Interspaced Short Palindromic Repeats). These techniques facilitate targeted modification of any crop genome sequence to generate novel variation for giving impetus to plant breeding programs (Scheben et al., 2017).

Comparative genomics and proteomics, pathogenesis-induced proteins, resistance/tolerance proteins, functional genomics, and metabolomics juxtaposed with plant-pest models provides better idea on account of host–pathogen interactions (Donatelli et al., 2017). New plant-specific comparative genomic databases and novel methods used to create them are helpful in integrating functional data into these databases the prominent ones including GOLD (Genome online Database, NCBI genomes, CoCrepedia andpla Bi (Martinez, 2016; Fondong et al., 2016)).

Metagenomic studies on small interfering RNAs or siRNAs of plant genomes revealed the presence of sequences homologus to T-DNA of *Agrobacterium* (Kyndt et al., 2015). In many plants horizontal gene transfer has been found to be the basis of transfer and subsequent maintenance of tumor-inducing gene (T-DNAs) from *Agrobacterium* species into *Nicotiana*, *Linaria*, and *Ipomoea* species (Quisepe-Hnamanquispe et al., 2017; Kyndt et al., 2015). Horizontal gene transfer has been largely used from *Agrobacterium tumefaciens* and *Agrobacterium rhizogenes* into *Nicotiana* sp. and *Linaria* sp. (Chen et al., 2016; Lacroix and Citovsky, 2016). Evolutionary and phylogenomic analysis of molds and fungi have revealed the existence of lateral gene transfer and nearly one-half of genes in Trichoderma *pwcdCA-Zome* (41%) were obtained via lateral gene transfer from plants associated filamentous fungi of class Ascomycota (Druzhinina et al., 2018).

The investigation of plant pathogen effector repertoire is possible with available genomics techniques in order to generate useful information about durable resistance (Gibriel et al., 2016). Crop breeding focuses on development of disease-resistant genes especially *xop J4* and *avrBST* for resistance against bacterial pathogens (Timilsina et al., 2016). Draft sequencing of pathogenic genomes provides a rapid and cost effective means to understand life and disease cycle attributes (Muller et al., 2016; Xiang et al., 2016).

In case of crops with large genomes plant breeders depend on genotyping-by-sequencing (GBS) to map quantitative traits loci for tolerance to fungal infections (Savary, 2014). Pariyar et al. (2016) has described use of association mapping for identifying association between phenotypic variations in susceptibility to nematode infections. Single Nucleotide Polymorphisms studies have contributed to plant-pathogen interaction for bacterial, viral, and fungal infections (Gao et al., 2016). Role of Leucine-rich repeat LRR-receptor like kinase and NBS-LRR genesin disease resistance has been elucidated through Linkage Disequilibrium Analysis by SNPs (Chang et al., 2016).

13.7 DISEASE DETECTION METHODS

Biotechnology based direct methods of plant disease detection such as polymerase chain reaction (Vincelli, 2016), fluorescence in situ hybridization (Kliot et al., 2014), enzyme-linked immunosorbent assay (Fagalawa et al., 2013), flow cytometry (Davidson et al., 2012), and in addition gas chromatography-mass spectrometry (Sparkman et al., 2011) have played a significant role in plant disease management. In addition, infrared thermography (Mahlein, 2016), fluorescence imaging (Bauriegel and Herppich, 2014), hyperspectral techniques (Mahlein et al., 2012), and biosensors (Fang and Ramaswamy, 2015) are some indirect methods for detecting plant diseases.

13.8 NUCLEIC ACID-BASED METHODS

Molecular methods based on DNA/RNA probe technology includes southern hybridization, in situ hybridization, fluorescence in situ hybridization (Kliot et al., 2014) microarray (Krawczyk et al., 2017; Tiberini and Barba, 2012), and macroarray (Lievens et al., 2012), isothermal amplification technology including Loop-mediated isothermal amplification (Yasuhasa et al., 2016; Fischbach et al., 2017), Strand displacement amplification (Walker et al., 1992), Nucleic acid sequence-based amplification (Compton, 1991), Recombinase polymerase amplification (Piepenburg et al., 2016), Helicase dependent amplification (Vincent et al., 2004), Hinge-Initiated Primer Dependent Amplification of Nucleic Acids (Fischbach et al., 2017), DNA barcoding (Ray et al., 2017), and so forth for detection of pathogens and infections. Currently, biosensors are used for early detection with high sensitivity and specificity for plant pathogens based on both antibody and DNA receptors. Nanotechnology based nanochannels and metallic nanoparticles have been helpful in management of bacterial and viral infections in crops (Khater et al., 2017).

A method widely gaining popularity based on massive parallel transcriptome-wide sequencing of RNA is RNA-Seq, which conducts RNA editing and splicing, sequence mapping, transcriptome reconstruction, polymorphisms to understand allele-specific expression, and quantification of individual transcript isoforms (Gautam et al., 2017). RNA interference (RNAi) is an RNA silencing-based powerful and novel gene therapy against fungal, viral, and bacterial infection in plants based on gene expression regulation through mRNA degradation, translation repression, and chromatin remodeling with the aid of small noncoding RNAs (ISAAA, 2018). The process of silencing is based on products of the double-stranded RNA called small interfering RNAs, also reported to play a pivotal role in functional genomics (Silva et al., 2018) in terms of transcriptional and posttranscriptional gene silencing, during complex plant-pathogen interaction through production of snciRNAs by both plants and pathogens (Silva et al., 2018).

13.9 MANIFESTATION OF THE OMICSERA

The organizations and functions of biological systems work are well comprehended through genomics, transcriptomics, proteomics, and metabolomics along with bioinformatics based on dissemination of information and knowledge in terms of physiological, pathological, and environmental molecular variability. Sophisticated and cutting-edge technology can be used to plan judicious utilization of resources for agriculture and the subsequent management of plant diseases in great details, precision, and accuracy (Schoonbeek et al., 2016).

13.10 GENOMICS

Genomics has acted as a versatile tool in understanding molecular mechanisms of plant disease and hence their management. Engineered plant resistance to pathogens based on genes derived from different pathogenic and nonpathogenic organisms, genetic editing, and functional genomics have been instrumental in this aspect (Fondong et al., 2016; Jain, 2012). Based on artificial miRNA, trans-acting small interfering RNA and some CRISPR associated protein 9 (Cas 9) technology and also targeting induced local lesions in genomes a number of studies have been carried out (Fondong et al., 2016). On a broader scale molecular markers, genomics and NGS technologies (references) have paved way for better understanding of pathogenesis and its remediation (Esposito et al., 2016). NGS technologies have played a major role in perpetuating genomics-based research in order to improve stress resistance in plants. A combination of whole-genome resources coupled with NGS technologies have helped in adding impetus to research on disease resistant crops by increasing accessibility and availability of complete plant and pathogen genome. Plant-microbe interaction is an established phenomena based on a precise communication system (Shelden et al., 2016). The intricacies of plant–microbe interaction are extensively studied using metagenomics for segregating beneficial and harmful microbes (Esposito et al., 2016) by generating plant/pathogen genome or transcriptome marker sequences of virulence phenotypes in pathogen or resistance phenotypes in the plants leading to improvement in plant disease management (Burdon et al., 2016). A number of NGS technologies Roche 454 (http://454.com/products/gs-flx-system/), Illumina HiSeq Specifications (http://www.illumina.com/systems/hiseq-3000-4000/html), PacBioRS11 (http://files.pacb.com/pdf/PacBio-RS-11-Brochure.pdf) have facilitated a revolution in field of nucleic acid sequencing in terms of efficiency and output (Esposito et al., 2016).

13.11 METABOLOMICS

Various metabolites in different proportions are produced by plants with a specific role in overall plant growth, stepwise development, and appropriate conditional response (Afendi et al., 2012; Hong et al., 2016). Integrated technologies such as nuclear magnetic spectroscopy, mass spectrometry based methods such as gas chromatography-MS, Liquid chromatography-MS, Capillary electrophoresis-MS, and Fourier transform ion cyclotron resonance-MS have played a major role in designing metabolic profiles of plants and infected plants (Matsuda et al., 2010; Kumar et al., 2017).

Exhaustive data set generated from high-throughput tools are processed through data processing platforms such as MET-COFEA, Met-Align, Chroma-TOF, and MET-XAlign (Misra and Vander Hooft, 2016; De Souza et al., 2017). Baseline-correction, alignment, deconvolution, normalization, and so forth prior to compound identification. Metabolome databases such as METLIN, NIST, GOLM, and so forth are used for metabolite identification (Johnson and Lange, 2015; Kumar et al., 2017). Identified data are statistically analyzed with correlation maps, principal component analysis, partial least squares, and so forth and using web tools and software's such as MetaboAnalyst, Cytoscape, and so forth (Xie et al., 2015; Tsugawa et al., 2015).

Metabolomics integrated with transcriptomic analyzes has been instrumental for investigating metabolic dynamics in terms of molecular mechanisms underlying behavior toward different stresses in fruits and cereal crops (Venkatesh et al., 2016; Sheldon et al., 2016; Hong et al., 2016). Functional genomics studies based on metabolomics combined with Quantitative Trait Locus, Genome-Wide Association Studies and Knockout/Knockdown technology reverse genetics, RNAi, and gene knockout has elucidated the biochemistry of plants in stress (Cho et al., 2016; Wen et al., 2016)

13.12 TRANSCRIPTOMICS AND PROTEOMICS

Gene-protein and biomolecules based studies in phytopathogen systems have been conducted using transcriptome shotgun sequencing (RNA-Seq) in addition to shotgun nanoflow scale liquid chromatography–tandem mass spectrometry (LC–MS/MS) proteomics, which have thrown light on behavior and response of above and underground plant tissue during infection (Schenk et al. 2012; Seco-Garcia et al., 2017). Whole-genome transcriptome and proteomics analysis of infection causing pathogens have revealed roles of diverse genes in virulence such as those coding for enzymes, peroxidises, cytochrome P-450, effector proteins, transcription factor genes, and genes involved in resistance response (Bengtsson et al., 2014).

Manifold advances in transcriptomics have made RNA-sequencing (RNA-seq) an important biotechnological tool for studying gene expression under stress (Gautam et al., 2017). Among phytopathogenic bacteria members of genus *Xanthomonas* are found all over the world affecting a large variety of monocots and dicots especially *Xanthomonas citri pv. Citri*, *Xanthomonastranslucens*, and *Xanthomonas campestris*. Studies related to changes in proteome of plants as a result of pathogenic inoculation have been conducted in wheat, rice, barley, and so forth by combining whole transcriptome shotgun sequencing (RNA-seq) with shotgun nanoflow scale LC-MS/MS proteomics in order to study plant response to infection (Seco et al., 2017). Effects of calcium ions on SA-induced resistance for *Botrytis cinerea* in tomato have been analyzed through proteomics analysis to uncover inducible proteins in tomato (Linlin et al., 2016).

Proteomics allows us to segregate protein complement or proteome, further analyzed by high resolution of 2-dimensional electrophoresis (2DE) and mass spectrometry and finally de novo sequencing ultimately leading to protein identification (Meng et al., 2018). Changes in enzymatic activity and protein expression in antioxidative system within leaf apoplast of *Prunus persica* c.v. GS3O5 (peach) on plum pox potyvirus (PPV infection) (Diaz Vivancos et al., 2006a, 2006b). 2DE and subsequent MALDI-TOFMS have been performed to analyze induced expression of nuclear proteins in *Capsicum annum cv. Bugang* (hot pepper) infected by TMV (Lee et al., 2006; Xu et al., 2011).

Using proteomics, the proteosome subunits such as 26S proteosome subunit RPN7, Ubiquitin extension proteins, and so forth have been yielded from *C.annum*, Thaumatin-like protein, and Mandilonitrile lyase isoform MDL5 precursor from *Prunus sp.* during viral infections using proteomics (Mehta et al., 2008), Type III secretion system (TTSS) of pathogenic bacteria such as *Xanthomonas*, *Psuedomonas*, and so forth have yielded sulpfate binding proteins, inorganic pyrophosphatase, Cellulase, Metalloprotease A, and so forth by proteomics analysis (Goldy et al., 2017).

Proteomics studies have also reported global protein expression and reference maps of important bacterial pathogens including *Xanthomonas fastidiosa* (Smolka et al., 2003) and *Agrobacterium tumefaciens* (Rosen et al., 2004). Mehta et al. (2008) analyzed proteomics and transcriptomic profiles of *Arabidopsis thaliana* leaves during early responses to the challenge by *Psuedomonas syringae* pv. Tomato. When pathogenic fungi start infection process secreted and intracellular proteins are upregulated or downregulated, improving predation ability of fungus (Murad et al., 2006, 2007). A number of studies have helped in understanding dimorphic transition from budding to filamentous growth as well as appressorium construction (Grenville-Briggs et al., 2005). Fungal infections by *Phytopthora infestans*, *Fusarium graminearum*, *Phytopthora ramorum*, and so forth have been analyzed by proteomicsformethionine synthase (Pi-met 1) gene, Threonine synthase Chitinase, Ceramidase, and so forth (Phalip et al., 2005; Meijer et al., 2006).

Obligatorysedentary endoparasites among nematodes such as those of *Meliodogyne spp.*, *Heterodera* spp., and *Globodera* spp. attack plants in larvae and adult form, damaging the root system, which leads to extreme deficiency of water and nutrients (Chitwood, 2003). A large number of nematode EST libraries have been designed and a number of proteins have been identified such as Chaperonin-HSP 60, Enolase, Calreticulin, and so forth (Curtis, 2007; Gao et al., 2003). In nematode pathogenesis, 2D-electrophoresis allied to MS has been used to rapidly generate peptide sequence tags, which can be linked to ESTs in silico (Mehta et al., 2008). In addition, EST libraries obtained by microaspiration of cytoplasmic material from esophageal glands of *M. incognita* and *H. glycines* have revealed an array of unknown proteins (Gao et al., 2003; Huang et al., 2003). A number of species of problematic cereal nematode pests have been identified and phylogenetically analyzed using internal transcribed spacer and species specific PCR (Jiang-kuan et al., 2017). Using proteomics approach, three proteins have been identified in response to nematode infection including chitinase and a PR protein in *Coffea canephora* and a quinine reductase 2 in *Gossypium hirsutum* (Mehta et al., 2008).

Ubiquitin has been identified as key player in plant developmental and stress response and on these lines Ubiquitin-Protease System (UPS) has been found to posttranslationally modify cellular proteins with small amount of ubiquitin causing regulated degradation by proteasome (Adams and Spoel, 2018). Hormone responsive gene expression profiles, including the ones triggered by immune hormone salicylic acid (SA) have been observed for their behavior under UPS system. Components of UPS pathway have been reported to be utilized by SA for reprogramming of transcriptome in order to establish local and systemic immunity. SA induces the activity of Cullin RING ligases (CRLs), which fuse the chains of ubiquitin to downstream transcriptional regulator and as a result target them for degradation by the proteosome (Adams and Spoel, 2018).

13.13 BIOINFORMATICS

Bioinformatics has become an inevitable tool for an in-depth research on plant disease management based on its ability to elucidate the multifaceted aspects and functionalities of existing biomolecules and possible novelties in many aspects of agricultural sciences (Esposito et al., 2016). It is a systematic amalgamation and application of information technology and biological information by addressing data collection and warehousing, data mining, database searches, analyzes and interpretation, modeling, and product design using bioinformatics tools and techniques. Bioinformatics has been helpful in decoding plant genomes based on sequencing in model organisms and application of high throughput experimental methods in addition to imparting biological research an in silico dimension (Singh et al., 2011). Bioinformatics based study of host–pathogen interactions, disease genetics, pathogenicity, and disease resistance based on gene-protein functional dynamics has contributed substantially for improvised plant disease management (Jayaram and Dhingra, 2010).

Plant disease management in terms of resistance is most crucial for plant breeding programs in which phenotypic and pedigree information for agronomic and resistance traits are integrated (Vassilev et al., 2005, 2006). Improved algorithm and computing power stimulate and optimize selection strategies as well as for studying pest-plant model based studies for epidomology (Michelmore, 2003). Bioinformatics helps in consolidation of all sequence data into public domain through repositories in order to provide rational annotation of genes, proteins. and phenotypes in terms of data on sequence information, information on mutations, markers, maps. and functional discoveries (Alemu, 2015).

A number of open-source software and reference database in genomics, transcriptomics, and metagenomics are used, which include FastQC, FASTX-toolkit (preprocessing), (META) VELVET/OASES, TRINITY (ASSEMBLY) Star, Tophat/Cufflinks (mapping) and Mothue, Qiime, and so forth (marker-based metagenome) (http://www.bioinformatics.babrahan.ac.uk/projects). In addition, Gene prediction/annotation task is executed by software named Ensembl genome annotation (Meta) Genemark, NCBI genome annotation tRNA scan-SE (http://www.ncbi.nlm.nih.gov/books/NBK1694391).

Web-based tools are an excellent platform to analyze huge data sets in terms of data storage, dissemination, and high throughput analysis such as CyVerse (Devisetty et al., 2016) and Galaxy (Devisetty et al., 2016; Bornich et al., 2016). CoGe (Lyons et al., 2008) is another platform for comparative genomics research, which has manifested into an open-ended network of interconnected tools for managing, analyzing, and visualizing next-generation sequencing data (https://genomeevolution.org/coge/). A few other tools include variant effector predictor (McLaren et al., 2016) provided by Ensembl and Gramene, Pathway enrichment and comparison tools by Plant Reactome (Naithani et al., 2016).

13.14 FUTURE PROSPECTS

A number of biotechnological tools have enabled us to carry an in-depth analysis of plant pathogenesis in terms of nature, causes, and mitigation.

Various pathogens and their influence on crops and plants have been identified and characterized with the help of cutting-edge biotechnology to facilitate the vast science of plant disease/pathology and its management. Biotechnology has enabled us to understand the high genetic variability of different pathogens of various crop plants and their genetic expression and regulation, spatially, and temporally. The fundamental genetic variations among the pathogens have been comprehended at molecular level with advanced fields of proteomics, transcriptomics, and metabolomics, and the entire information has been converted into sophisticated and accessible databases using bioinformatics tools. Biotechnology thus is an integral part of any agricultural research study and the applied aspects of this science can prove beneficial in the long run upon careful and prioritized consideration and requirement.

KEYWORDS

- **pathogen**
- **disease**
- **management**
- **biotechnological approaches**

REFERENCES

Adams, E. H. G and Spoel, S. H. (2018). The Ubiquitin–proteaseome system as a transcriptional regulator of plant immunity. *J. Exp. Botany.* 2016. http://doi.org/10.1093/jxb/ery216.

Afendi, F. M., Okada, T., Yamazaki, M., Hirai-Morita, A., Nakamura, Y., Nakamura, K., Nakamura, Y, Ikeda, S., Takahashi, H., Altaf-Ul –Amin, M., Darusmnan, L. K., Saito, K. and Kanaya, S. (2012). KNApSAcK family databases: integrated metabolite-plant species databases for multifaceted plant research. *Plant Cell Physiol.* 53. http://doi.org/10.1093/pcp/pcr165.

Agrios, G. N. (2005). *Plant Pathology*. Fifth Edition. New York: Academic Press. p. 633.

Alexandratos, N and Bruinsma, J. (2012). *World Agriculture towards 2030/2050 FAO*, 12th Revision. Rome: FOA.

Arlat, M., Van Gijsegen, F., Huet, J. C., Pernollet, J. C. and Boucher, C. A. (1994). PopA1, a protein which induces a hypersensitivity like response on specific Petunia genotypes, is secreted via the Hrp pathway of *Psuedomonas solanacearum. EMBOJ.* 13: 543–553.

Ashry, N. A., Ghonaimn, M. M. Mohammed, H. I. and Mogazy, A. M. (2018). Physiological and molecular genetic studies on two elicitors for physiological and molecular genetic studies on two elicitors for soyabean cultivars improving the tolerance of six Egyptian to cotton leafworm. *Plant Phys. Biochem.* 130: 224–234.

Azhaguvel, P., Vidya, D Sharma, A and Varshney, R. K. (2006). Methodological advancement in molecular markers to delimit the gene (s) for crop improvement. In; (Ed. Texeira de Silva J.). *Floriculture Ornamental and Plant Biotechnology Advances and Topical Issues.* London, U.K: Global Science Books. pp: 460–469.

Bai, Y. S., Pavan, Z., Zheng, N. F., Zappel, A., Reinstadler, C., Lotti, C., De Giovanni, L., Ricciardi, P., Lindhout, R., Visser, Theres and Panstruga, (2008). K. R. Naturally occurring broad-spectrum powdery mildew resistance in a Central American tomato accession is caused by loss of mlo function. *Mol. Plant Microbe Interact.* 21: 30–39.

Barbieri, L., Valbonesi, P., Bonora, E., Gorini, P., Bolognesi, A., Stirpe, F. (1997). Polynucleotide: adenosine glycosidase activity of ribosome-inactivating proteins: effect on DNA, RNA and poly(A). *Nucleic Acids Res.* 25: 518–522.

Bauriegel, E and Herppich, W. B. (2014). hyperspectral and chlorophyll fluorescence imaging for early detectio of plant diseases with special reference to fusarium spec. infections on wheat. *Agriculture.* 4 (1): 32–57.

Beljah, K., Chaparro, A., Kamoun, S., Petersn, N. and Nekrasov, V. (2015). Editing plant genomes with CRISPR-Cas9. *Current Opinion Biotechnol.* 32: 76–84.

Bengtsson, T., Weighill, D., Prox-Wera, E., Levande,r F., Resji, S., Burrer, D. D., Moushib, L. I., Hedley, P. E., Liljeroth, E., Jacobson, D., Alexandersson, E and Andreasson, E. (2014). Proteomics and transcriptomics of the BABA-induced resistance response in potato using a novel functional annotation approach. *BMC Genomics.* 15: 315.

Bhagat, S. (2010). *In vitro Efficacy of Burma Dhania (Eryngium foetidum) against Soil Borne and Foliar Plant Pathogens of Tomato.* Port Blair : IARI annual report (2010-11),.

Bhagat, S., Birah, A., Kumar, R., Yadav, M. S. and Chatopadhyaya, C. (2014). Plant disease management. prospects of pesticides of plant origin. *Adv. Plant Biopest.* http://doi.org/10.1007/978-81-322-20006-0-7.

Bhattarai, K. K., Li, Q., Liu, Y., Dinesh-Kumar, S.P., Kaloshian, I. (2007). The Mi--mediated pest resistance requires Hsp90 and Sgt1. *Plant Physiol.* 144: 312–323.

Bornich, C., Grytten, I., Hovig, E., Paulsen, J., Cech, M., Sandve, G. K. (2016). Galaxy Portal: Interacting with the galaxy platform through mobile devices. *Bioinformatics.* 32(11): 1743–1745.

Boutrot, F. and Zipfel, C. (2017). Function, discovery, and exploitation of plant pattern recognition receptors for broad-spectrum disease resistance, *Annu. Rev. Phytopathol.* 55: 257–286.

Boxall, A. B. A., Hardy, A., Beulke, S., Boucard, T., Burgin, L., Falloon, P. D., Haggarth, P. M., Hutchinson, T., Kovatsari, R., Leonardo, G., Levy, S. L., Nicholas, G., Parsons, S. A., Potts, L., Stone, D., Topp, E., Turley, D. B., Walsh, K., Wellington, E. M. H. and Williams, R. J. (2009). Impacts of climate change on indirect human exposure to pathogens and chemicals from agriculture. *Environ. Health Perspect.* 117: 508–514.

Bradford, J. K., Dahal, P., Asbrown, J. V., Kunusoth, K., Bello, P., Thompson, J. and Wu, F. (2018). The dry chain Reducing post harvest losses and improving food safety in humid climate. *Trends Food Sci. Technol.* 71: 84–93.

Brown, L. R. (2011). *World on the Edge. How to Prevent Environmental and Economic Collapse.* New York, London W.W: Norton and Company.

Burdon, J. J., Zhan, J., Barrett, L.G., Papain, J. and Thrall, P. H. (2016). Addressing the challenges of pathogen evolution on the world's arable crops. *Phytopathology.* 106: 1117–1127.

Buttner, D. (2016). Behind the lines-actions of bacterial type III effector proteins in plant cells. *FEMS Microbiol. Rev.* 40: 894–937.

Callot, C., and Gallois, J.L. (2014). Pyramiding resistances based on translation initiation factorsin Arabidopsis is impaired by male gametophyte lethality. *Plant Signal. Behav.* 9: e27940.

Campos, A., Silva, M. S., Magalhaes, C. P., Ribeiro, S. G., Sarto, R. P., Vieira, E. A., Grossi-de-Sa, M. F. (2008). Expression in Escherichia coli purification, refolding and antifungalactivity of an osmotin from Solanum nigrum. *Microb. Cell Factories.* 7: 1–10.

Candido, E. S., Cardoso, M. H., Sousa, D.A., Romero, K. C., Franco, O. L. (2015). Proteinaceous plant toxins with antimicrobial and antitumor activities. In: (Eds. Gopalakrishnakone, P., Carlini, C.R., Ligabue-Braun, R.), *Plant Toxins*. Netherlands, Dordrecht: Springer,. pp. 1–14.

Candido, E. S., Pinto, M. F., Pelegrini, P. B., Lima, T. B., Silva, O.N., Pogue, R., Grossi-de-Sa, M. F. and Franco, O. L. (2009). Plant storage proteins with antimicrobial activity: novel insights into plant defense mechanisms. *FASEB J.* 25: 3290–3305.

Charles, H., Godfray, J., Beddington, J. R., Crute, I. R., Haddad, L., Lawrence, D., Muir, J. F., Pretty, J., Robinson, S., Thomas, S. M. and Toulmin, C. (2010). Food Security, The challenge of feeding 9 billion people. 327(5967): 812–818.

Chen, K., Borne, F.D., Julio, E., Obs zynski, J., Pale, P. and Otten, L. (2016). Root specific expression of opine genes and opine accumulation in some cultivars of the naturally occurring genetically modified organisms, *Nicotiana tabacum. Plant J.* 87: 258–269.

Chisholm, S. T., Coaker, G., Day, B. and Staskawicz, B. J. (2006). Host-microbe interactions: shaping the evolution of the plant immune response. *Cell.* 124: 803–814.

Chitwood, D. J. (2003). Research on plant parasitic nematode biology conducted by the US Department of Agriculture Agricultural Research Services. *Pest Manag. Sci.* 59: 748–753.

Cho, K., Cho, K., Sohn, H., Ha, I. J. and Hong, S, (2016). Network analysis of the metabolome and transcriptome reveals novel regulation of potato pigmentation. *J. Exp. Bot.* 67: 1519–1533.

Coll, N. S., Epple, P. and Dangl, J. L. (2011). Programmed cell death in the plant immune system. *Cell Death Differ.* 18: 1247–1256.

Compton, J. (1991). Nucleic acid sequence-based amplification. *Nature.* 350: 91–92.

Curtis, R. H. (2007). Plant parasitic nematode proteins and the host-parasite interaction. *Brief. Funct. GenomicProteomic.* 6: 50–58.

Dangi, J. L., Horvath, D. M. and Staskawicz, B. J. (2013). Pivoting the plant immune system from dissection to deployment. *Science.* 341(6147): 746–751.

Dangi, J. L. and Jones, J. D. (2001). Plant pathogens and integrated defence responses to infection. *Nature.* 411: 826–833.

Davidson, B., Dong, H. P., Berner, A. and Rieberg, B. (2012). The diagnostic and research application of flow cytometry in cytopathology. *Diagn. Cytopathol.* 40(6): 625–635.

De Ronde, D., Butterbach, P. Kormelink, R. (2014). Dominant resistance against plant viruses, *Front. Plant Sci.* 5: 307.

Deshpande, T. (2017). *A Report entitled State of Agriculture in India by PRS (An Independent NoPprofit Group*).

De Souza, L. P., Naake, T., Tohge, T. and Fernie, A. R. (2017). From chromatogram to analyte to metabolite. How to pick horses for courses from the massive web-resources for mass spectral plant metabolomics. *Gigascience*. https://doi.org/10.1093/gigascience/gix037.

Devisetty, U. K., Kennedy, K., Sarando, P., Merchant, N. and Lyons, E. (2016). Bringing your tools to CyVerse discovery environment using docker. *F1000Res.* 5: 1442.

De Wolf, E. D. and Isard, S. A. (2007). Disease cycle approach to plant disease prediction. *Ann. Rev. Phytopatholol.* 45: 203–220.

Diaz-Vivancos, P., Rubio Mesonero, V., Periago, P. M., Barcelo, A. R., Martinez-Gomez, P. and Hernande, J. A. (2006a). The apoplastic antioxidant system in Prunus: response to long term plum pox virus infection. *J. Exp. Botany.* 57: 3813–3824.

Diaz-Vivancos, P., Rubio, M., Mesonero, V., Periago, P. M., RoBeucelo, A., Martinez-Gomez, P. and Hernandez, J. A. (2006b). The apoplastic antioxidant system in Prunus response to long term plum pox virus infection. *J. Exp. Bot.* 57(14): 3813–3824.

doi: 10.1128/m Bio.00863-16

DOI:10.1080/15592324.2018.1454816

Donatelli, M., Magarey, R. D., Bregaglio, S., Willocquet, L., Whish, J. P. M. and Savary, S. (2017). Modelling the impacts of pest and diseases on agricultural systems. *Agric. Syst.* 155: 213–224.

Druzhinina, I. S., Chenthamara, K., Zhang, J., Atanasova, L., Yang, D., (2018). Massive lateral transfer of genus encoding plant cell wall degrading enzymes to the mycoparasitic fungus Trichoderma from its plant associated hosts. *PLOS Genet.* 14(14): e1007322. https://doi. org/10. 1371/journal.pgen.1007322.

Esposito, A., Colantuono, C., Ruggiers, V. and Cuisano, M. L. (2016). Bioinformatics for agriculture in the Next Generation Sequencing era. *Chem. Biol. Technol. Agric.* 3: 9.

Fagalawa, L. D., Kutama, A. S. and Yakasai, M. T. (2013). Current issues in plant disease control: Biotechnology and plant disease. *Bayero J. Pure Appl. Sci.* 6(2): 121–126.

Fang, Y. and Ramaswamy, R. P. (2015). Current and prospective methods for plant disease detection. *Biosensors.* 4: 537–561.

Fischbach, J., Frohme, M. and Glökler, J. (2017). Hinge-initiated primer-dependent amplification of nucleic acids (HIP). A new versatile isothermal amplification method. *Sci. Rep.* 7: 7683.

Fondong, V. N., Nagalakshmi, U. and Dinesh Kumar (2016). Novel functional genomics approaches. A promising future in the combat against plant virus. *Phytopathology.* 106: 1231–1239.

Fradin, E. F., Abd-El-Haliem, A., Masini, L., Van den Berg, G. C., Joosten, M. H. and Thomma, B. P. (2011). Interfamily transfer of tomato Ve1 mediates Verticillium resistancein Arabidopsis. *Plant Physiol.* 156: 2255–2265.

Gao, B., Allen, R., Maeir, T., Davis, E. l., Baum, T. J. and Hussey, R. S. (2003). The parasitome of the phytonematode *Heterodera glycines. Mol. Plant Microbe Interact.* 16: 720–726.

Gao, Q., Zhu, S., Kachroo, P. and Kachroo, A. (2015). Signal regulators of systemic acquired resistance. *Front. Plant Sci.* 6: 228.

Gautam, P. C., Shah, T. and Joshi, K. (2017). *Innovative Approach in Drug Discovery Ethnopharmacology. Systems Biology and Holistic Targeting.* New York: Academic Press. pp. 235–272.

Goldy, C., Svetaz, L. A., Bustamente, C. A., Allegrini, M., Valentini, G. H., Brinkovich, M. F., Fernie, A. R. and Lara, M. V. (2017). Comparative and proteomic and metabolomic studies between Prunus persica genotypes resistant and susceptible to Taphrina deformans suggest a molecular basis of resistance. *Plant Physiol. Biochem.*118: 245.

Grenville-Briggs, L. J., Avrova, A. O., Bruce, C. R., Williams, A., Whisson, S. C., Birch, P. R. and Van west, P. (2005). Elevated amino acid biosynthesis in *Phytopthora infestans* during appressorium formation and potato infection. *Fungal Genet. Biol.* 42: 244–256.

Gururani, M. A. and Park, S. W. (2012). Engineered resistance against filamentous pathogens in *Solanum tuberosum. J. Gen. Plant Pathol.* 78: 377–388.

Gururani, M. A., Venkatesh, J., Upadhyaya, C. P., Nookaraju, A., Pandey, S. K. and Park, S.W. (2012). Plant disease resistance genes: current status and future directions. *Physiol. Mol. Plant Pathol.* 78: 51–65.

Haggag, W. M. (2008). Biotechnological aspects of plant resistant for fungal diseases management. *Am. Eurasian J. Sustain. Agric.* 2(1): 1–18.

Hallwass, M., de Oliveira, A. S., de Campos Dianese, E., Lohuis, D., Boiteux, L. S., Inoue-Nagata, A. K., Resende, R. O. and Kormelink, R. (2014). The tomato spotted wilt virus cell to cell movement protein (NSM) triggers a hypersensitive response in Sw-5-containing resistant tomato lines and in *Nicotiana benthamiana* transformed with the functional Sw-5b resistance gene. *Mol. Plant Pathol.* 15: 871–880.

Hamamouch, N., Li, C., Seo, P. J., Park, C. M. and Davis, E. L. (2011). Expression of Arabidopsis pathogenesis related genes during nematode infection. *Molecular Plant Pathol.* 12(4): 355–364.

Hong, J., Yang, L., Zhang, D. and Shi, J. (2016). Plant metabolomics: an indispensable system biology tool for plant science. *Int. J. Mol. Sci.* 17: E767. https://doi.org/10.3390/ijms17060767.

Horvath, D. M., Stall, R. E., Jones, J. B., Pauly, M. H., Vallad, G. E., Dahlbeck, D. Staskawicz, B. J. and Scott, J. W. (2012). Transgenic resistance confers effective field level control of bacterial spot disease in tomato. *PLoS One*. 7: 1.

Huang, G., Gao, B., Maeir, T., Allen, R., Davis, E. L., Baum, T. J. and Hussey, R. S. (2003). A profile of putative parasitism genes expressed in the oesophageal gland cells of the root knot nematode Meliodogyne incognita. *Mol. Plant Microbe Interact.* 16 (5): 376–381.

ISAAA (2018). *Commercial G M Trait; Disease Resistance*,. http://www.isaaa.org/gmapprovaldatabase/commercial trait. default.asp 2 Trait Type 1D = 3 and Trait Type = Disease% 20 Resistance.

Jain, M. (2012). Next generation sequencing technologies for gene expression profiling in plants. *Brief Funct. Genomics*. 11: 63–70.

Jandu, J. J. B., Neto, R. N. M., Zagmignan, A., EMde Soursa, M. C. A., Brelazde-Castro, Dos Santos Correria, M. T. and da Silva, L. C. N. (2017). Targeting the immune system with plant lectinsto combat microbial infections. *Front Pharmacol.* 8: 671. https://doi.org/10.3389/fphar.2017.00671.

Jayaram, B and Dhingra, P., (2010). *Bioinformatics for a Better Tomorrow*, New Delhi: I.I.T, Hauz Khas.

Jiang-kuan, C. U. I., Peng, H., Shi-Ming, L., Erginbas Oracki, G., Wen-kun, H., Imren, M., Dababat A. A. and De-liang, P. (2017). Occurrence, identification and phylogenetic analysis of cereal cyst nematodes (*Heterodera* spp.) in Turkey. *J. Integ. Agr.* 16 (8): 1767–1776.

Johnson, S. R. and Lange, B. M. (2015). Open-access metabolomics databases for natural product research: present capabilities and future potential. *Front. Bioeng. Biotechnol.* 3: 22.

Jones, J. D. and Dangi, J. L. (2006). The plant immune system, *Nature*. 444: 323–329.

Jose, J. and Usha, R. (2003). Bhendi yellow mosaic disease in India is caused by association of a DNA-Beta satellite with a begomovirus. *Virology.* 305(2): 310–317.

Kabak, B., Dobson, A. and Var, I. (2006). Strategies to prevent mycotoxin contamination of food and animal feed. A review. *Crit. Rev. Food Sci. Nutr.* 46(8): 593–619.

Kachroo, A., Vincelli, P. and Kachroo, P. (2017). Signaling mechanisms underlying resistance.

Kaiser, B., Vogg, G., Furst, U. B. and Albert, M. (2015). Parasitic plants of genus Cuscuta and their interaction with susceptible and resistant host plants. *Front. Plant Sci.* 4: 6–45.

Kamber, T., Pothier, J. F., Pelludat, C., Rezzonico, F., Duffy, B. and Smits, T. H. M. (2017). Role of the type VI secretion systems during disease interactions of Erwinia amylovora with its plant host. *BMC Genom.* 18: 628.

Khalid, A., Zhang, Q., Yasir, M. and Li, F. (2017). Small RNA based genetic engineering for plant viral resistance: application in crop protection. *Front. Microbiol.* 8: 43.

Khan, R. S., Darwish, N. A., Khattak, B., Ntui, V. O., Kong, Shimomae, K., K., Nakamura, I. and Mii, M. (2014). Retransformation of marker-free potato for enhanced resistance against fungal pathogens by pyramiding chitinase and wasabi defensin genes. *Mol. Biotechnol.* 56: 814–823.

Khater, M., Ade la E Muniz., and Merkocia, A. (2017). Biosensors for plant pathogen detection. Biosens. *Bioelectron.* 93; 72–86.

Kim, J. K., Jang, I. C., Wu, R., Zuo, W. N., Boston, R. S., Lee, Y. H., Ahn, I. P. and Nahm, B. H., (2003). Co-expression of a modified maize ribosome-inactivating protein and a rice basic chitinase gene in transgenic rice plants confers enhanced resistance to sheath blight. *Transgenic Res.* 12: 475–484.

Kliot, A., Kontsedalov, S., Lebedev, G., Brumin, M., Cathrin, P. B., Marubayashi, J. M., Skaljac, M., Belausov, E., Czosnek. H., and Ghanim, M. (2014). Fluorescence in situ hybridizations (FISH) for the localization of viruses and endosymbiotic bacteria in plant and insect tissues. *J. Vis. Exp.* 24 (84): e51030. https://doi.org/10.3791/51030.

Klosterman, S. J., Rollins, J. R., Sudarshana, M. R. and Vinatzar, B. A. (2016). Disease management in genomics era—Summaries of focus issue papers. *Phytopathology* 106: 1068–1070.

Krawczyk, K., Uszczyńska-Ratajczak, B., Majewska, A. and Borodynko-Filas, N. (2017). DNA microarray-based detection and identification of bacterial and viral pathogens of maize. *J. Plant Dis. Protection.* 124(6): 577–583.

Kumar, R., Bohra, A., Pandey, A. K., Pandey, M. K. and Kumar, A. (2017). Metabolomics for plant improvement: status and prospects. *Front. Plant Sci.* 07. https://doi.org/10.3389/fpls.2017.01302.

Kumar, D. and Kalita, P. (2017). Reducing post harvest losses during storage of grain crops to strengthen food security in developing countries. *Foods.* 6(1): 8.

Lacroix, B. and Citovsky, V. (2016). Transfer of DNA from bacteria to eukaryotes. *M. Bio.* 7(4). p1i: eoo863-16.

Lanzanova, C., Giuffrida, M. G., Motto, M., Baro, C., Donn, G., Hartings, H., Lupotto, E., Careri, M., Elviri, L. and Balconi, C. (2009). The Zea mays b-32 ribosome-inactivating protein efficiently inhibits growth of *Fusarium verticillioides* on leaf pieces in vitro. *Eur. J. Plant Pathol.* 124: 471–482.

Lawaju, B. R., Lawrence, K. S., Lawrence, G. W. and Klink, V. P. (2018). Harpin inducible defense signaling components impair infection by the ascomycete *Macrophomina phaseolina. Plant Physiol. Biochem.* 129: 331–348.

Lee, B. J., Kwon, S. J., Kim, S. K., Kim, K. J., Park, C. J., Kim, Y. J., Park, O. K. and Paek, K. H. (2006). Functional study of hot pepper 26 S proteasome subunit RPN7 induced by TMV from nuclear proteome analysis. *Biochem. Biophysics. Re. Commun.*351: 405–411.

Li, T., Liu, B., Spalding, M. H., Weeks, D. P.and Yang, B. (2012). High-efficiency TALEN-based gene editing produces disease-resistant rice. *Nat. Biotechnol.* 30: 390–392.

Lievens, B., Justé, A. and Willems, K. A. (2012). Fungal plant pathogen detection in plant and soil samples using DNA macro-arrays. *Methods Mol. Biol.* 835: 491–507.

Linlin, Li., Peng, G., Hua, J. and Tianlai, Li. (2016). Different proteomics of $Ca2^{+}$ on SA induced resistance to *Botrytis cinerea* in tomato. *Horticult. Plant J.* 2(3): 154–162.

Liu, N., Zhang, X., Sun, Y., Wang, P., Li, X., Pei, Y., Li, F. and Hou,Y. (2017a). Molecular evidence for the involvement of a poly galacturonase inhibiting protein, GhPGIP1, in enhanced resistance to *Verticillium* and *Fusarium* wilts in cotton, *Sci. Rep.* 7.

Liu, S. R,. Zhou, J. J., Hu, C. G., Wei, C. L. and Zhang Jin-Chi, (2017b). Micro RNA-mediated gene silencing in plant defense and viral counter defense. *Front Microbial.* 8: 1801.

Lopez-Cobello, R. M., Filppis, I., Bennett, M. H. and Turnbull, C. G. N. (2016). Comparative proteomics of cucurbit phloem indicates both unique and shared sets of proteins. *Plant J.* 88 (4): 633.

Ludke, D., Roth, C., Hartken, D and Weirmer, M. (2018). MOS6 and TN13 in plant immunity. *Plant Signaling Behav.* 13(4): e1454816.

Lyons, E., Pedersen, B., Kane, J., Alam, M., Ming, R., Tang, H., Wang, X., Bowers, J., Peterson, A., Lisch, D. and Freeling, M. (2008). Finding and comparing syntenic regions among *Arabidopsis* and the outgroups papaya, poplar and grape. CoGe with rosids. *Plant Physiol.* 148(4): 1772–1781.

Ma, W., Smigel, A., Tsai, Y. C., Braam, J. and Berkowitz, G. A. (2008). Innate immunity signaling:cytosolic Ca^{2+} elevation is linked to downstream nitric oxide generation throughthe action of calmodulin or a calmodulin-like protein. *Plant Physiol.* 148: 818–828.

Maag, D., Erb, M., Kollner, T. G. and Gershenzon, J. (2015). Defensive weapons and defense signals in plants: some metabolites serve both roles. *Bioessays.* 37: 167–174.

Mahlein, A. K. (2016). Plant disease detection by imaging sensors parallels and specific demands for precision agriculture and plant. *Phenotyping.* 100(2): 241–251.

Mahlein, A. K., Steiner, U., Hillnhutter, C., Deine, H.W. and Oerke, E. C. (2012). Hyperspectral imaging for small scale analysis of symptoms caused by different sugar beet diseases. *Plant Methods.* 8: 3.

Martin, J. A. and Wang, Z. (2011). Next generation transcriptome assembly. *Nat. Rev. Genetics.* 12: 671–682.

Matika, D. E. F. and Loake, G. J. (2014). Redox regulation in plant immune function, *Antioxid. Redox Signal. 21:* 1373–1388.

Matsuda, F., Hirai, M. Y., Sasaki, E., Akiyama, K., Yonekura-Sakakibara, K., Provart, N. J., Sakurai T, Shimada Y and Saito K, (2010). AtMetExpress development: a phytochemical atlas of Arabidopsis development. *Plant Physiol.* 152: 566–578.

McLaren, W., Gil, L, Hunt, S. E., Riat, H. A., Ritchie, G. R. S., Thormann, A., Flicek, P. and Cunninglame, F. (2016). The ensemble variant effect predictor. *Genome Biol.* 17 (1): 122.

Mehta, A., Brasileiro, A. C. M., Souza, D. S. L., Romano, E., Campos, M. A., Grossi-de Sa, M. F., Silva, M. S., Franco, O. L., Fragoso, R. R., Bevotori, R. and Rocha, T. L. A. (2008). Plant pathogen interaction. *FEBS J.* 275: 3731–3746.

Meijer, H. J., Van de, V. P. J., Yin, Q. Y., de Coster, C. G.and Klis, F. M. (2006). Grovers F and de Groot PW, 2006. Identification of cell wall associated proteins from *Phytopthora ramorum. Mol Plant Microbe Interact.* 19: 1348–1358.

Mendez, K. A. and Romera, H. M. (2017). Plant responses of pathogen attack. Molecular basis of qualitative resistance. *Rev. Fac. Nac. Agron.* 70(2): 8225–8235.

Mendoza, J. R., Kok, C. R., Stratton, J., Blanchini, A., Hallen-Adama, H. E. (2017). Understanding the mycobiota of maize from the highlands of Guatemala and implications for maize quality and safety. *Crop Protect.* 101: 5–11.

Mendoza, J. R., Sabillon, L., Martinez, W., Campabadal, C., Hallen Adams, H. E., Bianchini, A. (2016). Traditional maize post harvest management practices amongst small holders farmers in western highlands of Guatemala. *J. Stored Prod. Res.* 71: 14–21.

Meng, Q., Gupta, R., Min, C.W., Kim, J., Kramer, K., Wang, Y. Park, S. R., Finkemeier, I and Kim, S. T. (2018). A proteomic insight into the MSP1 and flg22 induced signalling in *Oryza sativa* leaves. *J. Proteomics.* http://doi.org/10.1016/jprot.2018.04.015.

Michelmore, R. W. (2003). The important zone :genomics and breeding for durable disease resistance. *Current Opinion Plant Biol.* 6: 397–404.

Misra, B. B., and van der Hooft, J. J. (2016). Updates in metabolomics tools and resources: 2014–2015. *Electrophoresis*. 37: 86–110.

Mugford, S. T., Qi, X., Bakht, S., Hill, L., Wegel, E., Hughes, R. K., Papadopoulou, K., Melton, R., Philo, M., Sainsbury, F., Lomonossoff, G. P., Roy, A. D., Goss, R. and Osbourn, J. A. A. (2009). Serine carboxypeptidase-like acyltransferase is required for synthesis of antimicrobial compounds and disease resistance in oats. *Plant Cell.* 21: 2473–2484.

Murad, A. M., Laumann, R. A., Lime Tde, A., Sarmento, R. B., Noronha, E. F., Rocha, T. L., Valadares-Inglis, M. C. and Franco, O. L. (2006). Screening of entomopathogenic *Metarhizium anisopliae* isolates and proteomic analysis of secretion synthesized in response to cowpea weevil (*Callosobruchus maculatus*) exoskeleton. *Comp. Biochem. Physiol. C. Toxicol. Pharmacol.* 142: 365–370.

Murad, A. M., Laumann, R. A., Mehta, A., Noronha, E. F. and Franco, O. L. (2007). Screening and secretomic analysis of entomopathogenic *Beauveria bassiana* isolates in response to cowpea weevil (*Callosobruchus maculatus*) exoskeleton. *Comp. Biochem. Physiol. C. Toxicol. Pharmacol.* 145: 333–338.

Musselman, L. J. and Press, M. C. (1995). Introduction to parasitic plants. In: (Eds. Press M.C., Graves J.D.) *Parasitic Plants*. Chapman and Hall, London. pp 1–13.

Muthamilarasan, M. and Prasad, M. (2013). Plant innate immunity: an updated insight into defense mechanism, *J. Biosci.* 38: 433–449.

Nadal, A., Montero, M., Company, N., Badosa, E.. Messeguer, J.. Montesinos, L.. Montesinos, E. and Pla, M. (2012). Constitutive expression of transgenes encoding derivatives of the synthetic antimicrobial peptide BP100: impact on rice host plant fitness, *BMC Plant Biol.* 12 (2012) 159–159.

Naithani, S. (2016). Plant Reactome: a resource for plant pathways and comparative analysis. *Nucleic Acids Res.* 45(D1): D1029–D1039. https://doi.org/10.1093/nar/gkw932.

Narusaka, M., Hatakeyama, K., Shirasu, K. and Narusaka, Y. (2014). Arabidopsis dual resistance proteins, both RPS4 and RRS1, are required for resistance to bacterial wilt in transgenic Brassica crops. *Plant Signal. Behav.* 9: 29130.

Nawrot, R., Barylski, J., Nowicki, G., Broniarczyk, J., Buchwald, W. and Goździcka-Józefiak, A. (2014). Plant antimicrobial peptides. *Folia Microbiol.* (Praha) 59: 181–196.

Ng, T. B., Wong, J. H. and Wang, H. (2010). Recent progress in research on ribosome inactivating proteins. *Curr. Protein Pept. Sci.* 11: 37–53.

Oard, S. V. and Enright, F. M. (2006). Expression of the antimicrobial peptidesin plants to control phytopathogenic bacteria and fungi. *Plant Cell Reports.* 25 (6): 561–72.

Oerke, E. C. (2006). Crop losses to pest. *J. Agic. Sci.* 144: 31–43.

Phalip, V., Delalande, F., Carapito, C., Goubet, F., Hatsch, D., Leize-Wagner, E., Dupree, P., Dorsselaer, A. V. and Jeltsch, J. M. (2005). Diversity of exoproteome of *Fusarium graminareaum* grown on plant cell wall. *Curr. Genet.* 48: 366–379.

Piasecka, A., Jedrzejczak-Rey, N. and Bednarek, P. (2015). Secondary metabolites in plant innate immunity: conserved function of divergent chemicals. *New Phytol. J.* 206: 948–964.

Prasad, K., Mathur, P. B., Waliyar, F. and Sharma, K. K. (2012). Over expression of chitinase gene in transgenic peanut confers enhanced resistance to major soil borne and foliar fungal pathogens. *J. Plant Biochem. Biotech.* 22(2): 222–233.

Prendeville, H. R., Tenhumberg, B and Pilson, D (2014). Effects of virus onplant fecundity and population dynamics. *New Phytol.* https://doi.org:10.1111/nph.12730.

Presti, L., Lanver, D., Schweizer, G., Tamaka, S., Liang, L., Tollot, M., Zuccaro, A., Reissmann, S. and Kahmann, R. (2015). Fungal effectors and susceptibility. *Annual Rev. Plant Biol.* 65: 513–545.

Quieter, F. (2016). The CRISPR-Cas 9 technology: closer to the ultimate tool kit for targeted genome editing. *Plant Sci.* 242: 65–76.

Okmen, B and Dochelmann, G, (2014). Inside plant; biotrophic strategies to modulate host immunity and metabolism. *Current Opinion Plant Biol.* 20: 19–25.

Quisepe-Huamanquispe, D. G., Gheysen, G and Kreuze, J. F. (2017). Horizontal gene transfer contributes to plant evolution: the case of agrobacterium T-DNAs. *Front Plant Sci.* 8: 2015.

Ray, M., Ray, A., Dash, S., Mishra A., Achary, K. G., Nayak, S., Singh, S. (2017). Biosens. *Bioelectron.* 87: 708–723.

Ren, T., Qu, F.and Morris, T. J. (2000). HRT gene function requires interaction between a NAC protein and viral capsid protein to confer resistance to turnip crinkle virus. *Plant Cell.* 12: 1917–1192.

Responses: what have we learned, and how Is It being applied? *Phytopathology*.107(12): 1452–1461.

Rivero, M., Furman, N., Mencaccia, N., Picca, P., Toum, L., Lentz, E., Bravo, F. and Mentaberry, A., (2011). Stacking of antimicrobial genes in potato transgenic plants confers increased resistance.to bacterial and fungal pathogens. *J. Biotechnol. 157*(2): 334–343.

Rosen, R, Sacher, A., Shehter, N., Becher, D., Buttner, K., Biran, D., Hecker, M. and Pon, E. Z. (2004). Two dimensional reference map of *Agrobacterium tumefaciens* proteins *Proteomics.* 4: 1061–1073.

Sankarana, S., Mishra, A., Ehsania, P. and Davis, B. (2010). A review of advanced technique for detecting plant diseases. *Compact Electron. Agric.* 72: 1–13.

Savary, S. (2014). The roots of crop health: Cropping practices and disease management. *Food Secur.* 6: 819–831.

Savary, S., Ficke A., Aubertot J. and Hollier, C. (2012a). Crop losses due to diseases and their implications for global food production losses and food security. *Food Secur.* 4: 519–537.

Savary, S., Ficke, A. and Auberlot, J. N. (2012b). Crop losses due to diseases and their implications for global food production losses and food security. In: *Food Security. The Science, Sociology and Economics of Food Production and Acess to Food.* Springer. ISSN 1876-4517.

Schenk, P. H., Carvalrais, L.C. and Kazon, K. (2012). Unravelling plant microbe interaction can multispecies transcriptomics help? *Trends Biotechnol.* 30: 177–184.

Schoonbeek, H. J., Wang, H. H., Stefanato, F. L., Craze, M., Bowden, S., Wallington, E., Zipfel, C. and Ridout, C. J. (2016). Arabidopsis EF-Tu receptor enhances bacterial disease resistancein transgenic wheat. *New Phytol. J.* 206(2): 606–613.

Schrot, J., Weng, A. and Melzig, M. F. (2015). Ribosome-inactivating and related proteins, *Toxins.* 7: 1556–1615.

Schuttelkopf, A. W., Gros, L., Blair, D. E., Frearson, J. A., Van Aalten, D. M. F. and Gilbert, I. H. (2010). Acetazolamide-based fungal chitinase inhibitors. *Bioorg. Med. Chem.* 18: 8334–8340.

Seco-Garcia, D., Chiapello, M., Bracale, M., Resce, C., Bagnaresi, P., Dubois, E., Moulin, L., Vannini C. and Koebnik, Q. R. (2017). Transcriptome and proteome analysis reveal new insight into proximal and distal responses of wheat to foliar infection by *Xanthomonas translucens*. *Sci. Rep.* 7: 10157.

Senthilkumar, R., Cheng, C. P. and Yeh, K.W. (2010). Genetically pyramiding protease-inhibitor genes for dual broad-spectrum resistance against insect and phytopathogens in transgenic tobacco. *Plant Biotechnol. J.* 8: 65–75.

Seo, J. K., Lii, J. Y., Li, Y. and Jin, H. (2013). Contribution of small RNA pathway components in plant immunity, *Mol. Plant Microbe Interact.* 26: 617–625.

Seo, Y. S., Rojas, M. R., Lee, J. Y., Lee, S. W., Jeon, J. S., Ronald, P., Lucas, W. J. and Gilbertson, R.L. (2006). A viral resistance gene from common bean functions across plant families and is up-regulated in a non-virus-specific manner. *Proc. Natl. Acad. Sci.* 103: 11856–11861.

Seybold, H., Trempel, F., Ranf, S., Scheel, D., Romeis, T., Lee, J. (2014). Ca2+ signalling in plant immune response: from pattern recognition receptors to Ca2+ decoding mechanisms. *New Phytol.* 204: 782–790.

Shah, J. and Zeier, J. (2013). Long-distance communication and signal amplification in systemic acquired resistance. *Front. Plant Sci.* 4: 30.

Shelden, M. C., Dias, D. A., Jayasinghe, N. S., Bacic, A. and Roessner, U. (2016). Root spatial metabolite profiling of two genotypes of barley (*Hordeum vulgare* L.) reveals differences in response to short-term salt stress. *J. Exp. Bot.* 67: 3731–3745.

Shi, Y., Liu, X., Fang, Y., Tian, Q., Jiang, H., Ma, H. (2018). 2,3 butanediol activated disease resistance of creeping bentgrass by inducing phytohormone and antioxidant response. *Plant Physiol. Biochem.* 129: 244–250.

Silva, M. S., Arraes, F. B. M., Camposc, M de A., De Sa, M. G., Fernandez, D. E. Cândido, De S. Cardoso, M. H., Franco, O. L. and De Sa, M. F. G. (2018). Review: Potential biotechnological assets related to plant immunitymodulation applicable in engineering disease-resistant crops. *Plant Sci.* 270: 72–84.

Silva, A. F., Matos, M. P., Ralph, M. T., Silva, D. L., deAlencar, N. M., Ramos, M. V. and Lima-Filho, J. V. (2016). Comparison of immunomodulatory properties of mannose binding lectins from *Canavalia brasiliensis* and *Cratylia aegentea* in a mice model of Salmonella infection. *Int. Immunopharmacol.* 31: 233–238.

Smolka, M. B., Martins, D., Winck, F. V., Santoro, C. E., Castellari, R. R., Ferrari, F., Brum, I. J., Galembeck, E., Della, C., Filho, M. and Machado, M. A.(2003). Proteome analysis of the plant pathogen *Xylella fastidiosa* reveals major cellular and extracellular proteins and a peculiar codon bias distribution. *Proteomics.* 3: 224–237.

Sole, M., Scheibner, F., Hartmann, A. K. N., Hause, G., Rother, A., Jordan, M., Lautier, M., Arlat M. and Buttner, D. (2015). *Xanthomonas campestris pv. Vesicatoria* secretes proteases and xylanases via the Xps Type II secretion system and outer membrane vesicles. *J. Bacteriol.* 197: 2879–2893.

Sparkman, O. D., Penton, Z. and Kitson, P. G. (2011). *Gas chromatography and Mass Spectrometry: A practical guide*. Academic Press. ISBN 978-0-08-092015-3. New York: Academic Press.

Stewart, L. R., Medina, V., Sudarshana, M. R.and Falk, B. W. (2009). Lettuce infectious yellow virus encoded P26 induces plasmalemma deposit of cytopathology. *Virology.* 338(1), 212–220.

Strange, R. N. and Scott, P. R. (2005). Plant disease: a threat to global food security. *Ann. Rev. Phytopathol.* 43: 83–116.

Tepfer, M. (2002). Risk assessment of virus resistant transgenic plants. *Annu. Rev. Phtopathol.* 40: 467–491.

Thurston, H. D. (1990). Plant-disease management—practices of traditional farmers. *Plant Dis.* 74: 96–102.

Tian, B., Yang, J. and Zhang, K. Q. (2007). Bacteria used in the biological control of plant parasitic nematodes: populations, mechanisms of action and future prospects. *FEMS Microbiol. Ecol.* 61(2): 197–213.

Tiberini, A. and Barba, M. (2012). Optimization and improvement of oligonucleotide microarray-based detection of tomato viruses and pospiviroids. *J. Virolog. Methods. 185*(1): 43–51.

Tiwari, S., Awasthi, H., Pandey, V. P. and Dwivedi, U. N. (2017). Genomics based approaches towards management of plant diseases with emphasis on in silico methods as a prudent approach. *J. Agr. Sci. Food Technol.* 3(3): 39–51.

Trivedi, P., Trivedi, C., Grinyer, J., Anderson, I. C. and Singh, B. K. (2016). harnessing host-vector microbiome for sustainable plant disease management of phloem-limited bacteria. *Front Plant Sci.* https://doi.org/10.3389/fpls.2016.01423.

Tsugawa, H., Cajka, T., Kind, T., Ma, Y., Higgins, B., Ikeda, K., Kanazawa, M, Vander Gheynst J, Fiehn O and Arita M. (2015). MS-DIAL: data-independent MS/MS deconvolution for comprehensive metabolome analysis. *Nat. Methods*. 12: 523 526.

Uma, B., Rani, T. S. Podile, A. R. (2011). Warriors at the gate that never sleep: non-host resistance in plants. *J. Plant Physiol.* 168: 2141–2152.

Van Bel, A. J. E. and Gaupels, F. (2004). Pathogen induced resistance and alarm signals in the phloem. *Mol. Plant Pathol.* (5): 495–504.

Van Loon, L. C. M. and Rep, C.M. (2006). Pieterse, Significance of inducible defense-relatedproteins in infected plants. *Annu. Rev. Phytopathol.* 44: 135–162.

Vassilev, D., Leunissen, J. A., Atanassov, A., Nenov, A. and Dimov, G. (2005). *Application of Bioinformatics in Plant Breeding*. Netherland: Wageningen University.

Vassilev, D., Nenov, A., Atanassov, A., Dimov, G. and Getov, L. (2006). Application of bio-informatics in fruit plant breeding. *J. Fruit Ornamental Plant Res.* 14: 145–162.

Venkatesh, T. V., Chassy, A. W., Fiehn, O., Flint-Garcia, S., Zeng, Q., Skogerson, K. and Harrigan, G. G. (2016). Metabolomic assessment of key maize resources: GC-MS and NMR profiling of grain from B73 hybrids of the nested association mapping (NAM) founders and of geographically diverse landraces. *J. Agric. Food Chem.* 64: 2162–2172.

Vincelli, P. (2016). Genetic engineering and sustainable crop disease management: opportunities for case-by-case decision-making. *Sustainability*. 8: 495.

Vincent, M., Xu, Y. and Kong, H. (2004). Helicase-dependent isothermal DNA amplification. *EMBO Rep.* 5: 795–800.

Walker, G. T., Fraiser, M. S., Schram, J. L., Little, M. C., Nadean, J. G. and Malinowski, D. P. (1992). Strand displacement amplification an isothermal in vitro DNA amplification technique. *Nucleic Acods Res.* 20 (7): 1691–1696

Weiberg, A., Wang, M. Bellinger, M. and Jin, H. (2014). Small RNAs: a new paradigm in plant microbeinteractions. *Annu. Rev. Phytopathol.* 52: 495–516.

Wen, W., Liu, H., Zhou, Y., Jin, M., Yang, N., Li, D., Luo, J., Xiao, Y., Pan, Q., Tohge, T., Fernie, A. R. and Yan. J. (2016). Combining quantitative genetics approaches with regulatory network analysis to dissect the complex metabolism of the Maize Kernel. *Plant Physiol.* 170: 136–146.

Xie, L. J., Chen, Q. F., Chen, M. X., Yu, L. J., Huang, L., Chen, L., Wang, F. Z., Xia, F. N., Zhu, T. R., Wu, J. X., Yin, J., Liao, B., Shi, J., Zang, J, Aharoni, A., Yao, N and Xiao, S, (2015). Unsaturation of very-long-chain ceramides protects plant from hypoxia-induced damages by modulating ethylene signaling in *Arabidopsis*. *PLoS Genet.* 11: e1005143. https://doi.org/10.1371/journal.pgen.1005143.

Xu, W., Yang, S., Bhadury, P., He, J., He, M., Gao, L., Hu, D. and Song, B. (2011). Synthesis and bioactivity of novel sulfone derivatives containing 2,4-dichlorophenyl substituted 1,3,4-oxadiazole/thiadiazole moiety as chitinase inhibitors. *Pestic. Biochem. Physiol.* 101: 6–15.

Yasuhasa, Bell, J. H., deSilva, A., Heuchelin, S. A., Chaky, J. L. and Alvarez, A. (2016). Detection of crosses with pathogen *Clavibacter michigenensis* subspecies *nebraskensis* in maize by loop-mediated amplification. *Phytopathology.* 106: 226–235.

Zhan, J., Thrall, P. H. and Burdon, J. J (2014). Achieving sustainable plantdisease management through evolutionary principals. *Trends Plant Sci.* 20: 1360–1385.

INDEX

C

M

Q

R

S

T

U

V

W

X

Y

Z

For Product Safety Concerns and Information please contact our EU representative GPSR@taylorandfrancis.com
Taylor & Francis Verlag GmbH, Kaufingerstraße 24, 80331 München, Germany

www.ingramcontent.com/pod-product-compliance
Lightning Source LLC
Chambersburg PA
CBHW060811220326
41598CB00022B/2590

* 9 7 8 1 7 7 4 6 3 9 4 3 6 *